SpringerBriefs in Mathematics

SpringerBriefs in Mathematics showcases expositions in all areas of mathematics and applied mathematics. Manuscripts presenting new results or a single new result in a classical field, new field, or an emerging topic, applications, or bridges between new results and already published works, are encouraged. The series is intended for mathematicians and applied mathematicians.

More information about this series at http://www.springer.com/series/10030

Taras Mel'nyk · Dmytro Sadovyi

Multiple-Scale Analysis of Boundary-Value Problems in Thick Multi-Level Junctions of Type 3:2:2

Taras Mel'nyk
Faculty of Mechanics and Mathematics
Taras Shevchenko National University
of Kyiv
Kyiv, Ukraine

Dmytro Sadovyi
Faculty of Mechanics and Mathematics
Taras Shevchenko National University
of Kyiv
Kyiv, Ukraine

ISSN 2191-8198   ISSN 2191-8201   (electronic)
SpringerBriefs in Mathematics
ISBN 978-3-030-35536-4   ISBN 978-3-030-35537-1   (eBook)
https://doi.org/10.1007/978-3-030-35537-1

Mathematics Subject Classification (2010): 35B27, 35B40, 35B20, 35J25, 36J61, 35J70, 35K20, 35K58, 74K30, 74Q05

This Springer imprint is published by the registered company Springer Nature Switzerland AG
The registered company address is: Gewerbestrasse 11, 6330 Cham, Switzerland

Preface

Multiple-scale analysis of boundary-value problems (elliptic and parabolic, linear and semilinear) in domains with very complicated structure is the main topic of this book. In the scientific literature such domains are called either *domains with very high oscillating boundaries* or *thick junctions*. Special kinds of such domains are considered in the book, namely thick multilevel junctions consisting of a cylinder on which thin annular discs from two different sets are ε-periodically and alternatively strung. The thin annular discs are divided into two sets (levels) depending on their geometric structure and boundary conditions imposed on their surfaces. Different Dirichlet, Neumann, and Robin (linear and nonlinear) boundary conditions are considered. The thin discs can have variable thickness vanishing at their edges, including the case when their boundaries are not Lipschitz.

Qualitatively different cases in the asymptotic behavior of solutions to these problems are discovered when the small parameter ε tends to zero, i.e., when the number of attached thin discs of each level infinitely increases and their thickness vanish. Different asymptotic methods and approaches developed intensively during the past two decades are used to prove convergence theorems, to construct asymptotic approximations for the solutions, and to derive the corresponding homogenized problems.

Results presented here have already been published in our articles in mathematical journals, but we give here a complete and unified presentation including a full literature review on this topic. The text will be useful for researchers and graduate students in asymptotic analysis and applied mathematics as well as for physicists, chemists, and engineers interested in the investigation of different processes, including heat and mass transfer, in their applications.

Kyiv, Ukraine
February 2019

Taras Mel'nyk
Dmytro Sadovyi

Acknowledgements The authors were partly supported by joint European grant EUMLS-FP7-People-2011-IRSES Project number 295164. The main authors' scientific activity and the preparation of the book have been carried out in the framework of the research project "Development of new analytic-geometric, asymptotic and qualitative methods for investigating invariant sets of differential equations" (number 19BF038-02) of the Taras Shevchenko National University of Kyiv.

Contents

Chapter 1
Introduction

Motivation

A lot of modern engineering constructions and biological systems have complex geometric forms. The reason is that some physical properties of materials are controlled by their geometric construction. Therefore, studying the effect of the geometric structure of a material can help to improve some of its beneficial physical properties and reduce undesirable effects. Mathematical models for this study are boundary-value problems (BVPs) in complex domains: perforated domains, grid-domains, thin domains with rapidly varying thickness, junctions of thin domains of different configurations, networks, etc.

A *thick junction* of the type $m : k : d$ is a union of some domain in $\mathbb{R}^n$ (called a *junction's body*) and a large number of thin domains, which are ε-periodically attached to some manifold on the boundary of the junction's body. This manifold is called a *joint zone*. The small parameter ε characterizes the distance between neighboring thin domains and also their thickness. The type $m : k : d$ of a thick junction refers, respectively, to limiting dimensions (as $\varepsilon \to 0$) of the junction's body (m), the joint zone (k), and each of the attached thin domains (d).

Various constructions shaped as thick junctions are successfully used in nanotechnologies, microtechnique, modern engineering constructions, for example, microstrip radiators, wide-bandgap semiconductors, efficient sensors (inertial, biological, chemical), signal processing filters, transistors, heat radiators [25, 40, 69, 71]. Figure 1.1 shows heat radiators in the form of thick junctions; such radiators are used in order to radiate more heat from a given total volume. Also, many biological systems (e.g., viruses, intestinal lining) have the thick junction form.

Now engineering approaches are capable of creating freestanding objects with villi thickness to about 20 nm. Therefore, it is often impossible to solve BVPs in thick junctions directly with numerical methods, because this would require too much CPU resources considering a large number of components of thick junctions

© The Author(s), under exclusive license to Springer Nature Switzerland AG 2019
T. Mel'nyk and D. Sadovyi, *Multiple-Scale Analysis of Boundary-Value Problems in Thick Multi-Level Junctions of Type 3:2:2*, SpringerBriefs in Mathematics,
https://doi.org/10.1007/978-3-030-35537-1_1

Fig. 1.1 Heat radiators shaped as thick junctions of the type 3:2:2

(in some cases few thousands). Increase in the size of computational domains for thick multi-structures naturally leads to longer computing times and makes it very difficult to maintain an acceptable level of the accuracy.

Thus, the development of new mathematical tools is necessary. One of them is asymptotic analysis of BVPs in thick junctions as $\varepsilon \to 0$, i.e., when the number of attached thin domains infinitely increases and their thickness decreases to zero. Asymptotic results give us the possibility to replace the original problem in a thick junction by the corresponding homogenized problem that is simpler and then apply computer simulation. In addition, in some cases it is possible to construct accurate and numerically implementable asymptotic approximations.

Also with the help of asymptotic analysis, it is possible to find new unexpected properties of solutions to such problems, which in turn make it possible to mathematically justify different physical (or biological) phenomena of such structures (see e.g., "Literature review" in this Introduction and Conclusions to Chapters).

The main difficulties in the study of BVPs in thick junctions

Thick junctions have a special character of the connectedness: there are points in a thick junction that are at a short distance of order $\mathscr{O}(\varepsilon)$, but length of all curves, which connect these points in the junction, is of order $\mathscr{O}(1)$. As a result, there is no sequence of extension operators $\{ E : W^{1,p}(D_\varepsilon) \mapsto W^{1,p}(\mathbb{R}^n) \}_{\varepsilon>0}$ whose norms are uniformly bounded in ε (see e.g. [15, 79]) ; here D_ε is some thick junction.

At the same time, the availability of a uniformly bounded family of extension operator is a typical supposition in overwhelming majority of the existing homogenization schemes for problems in perforated domains with the Neumann or Robin boundary conditions (see e.g., [59, 65, 73, 74]). The existence of such extensions for ε-periodically perforated domains was proved in [22, 124].

Also because of a such connectedness, it is impossible to apply the boundary trace embedding theorems in the Sobolev spaces $W^{1,p}(D_\varepsilon)$ with a constant that is independent of ε.

In addition, thick junctions are non-convex domains with non-smooth boundaries. Therefore, solutions of BVPs in such domains have only minimal H^1-smoothness, while the H^2-smoothness of solutions is necessary for some homogenization procedures (see e.g., [23]).

There are two ways to study BVPs in thick junctions. The first one consists of the proof of convergence results for solutions (so- called convergence theorems).

The second one is related to obtaining the formal asymptotics for a solution and its justification (establish an asymptotic error estimate). The corresponding homogenized problem is derived from the limit problems for each domain forming the thick junction with the help of solutions to junction-layer problems around the joint zone. However, junction-layer solutions have polynomial or logarithmic growth at infinity and do not decrease exponentially as in many boundary-layer problems. Their behavior depends on the junction type $m : k : d$, and they influence directly the leading terms of the asymptotics (see e.g., Chap. 5). This explains the introduction of such a classification.

It should be noted here that an important problem for each asymptotic method is its accuracy. Therefore, the proof of the error estimate for the discrepancy between the constructed approximation and the original solution is a general principle that has been applied to the analysis of the efficiency of the proposed asymptotic method. How to do this, if the junction-layer solutions grow at infinity was until recently an open question.

All those factors mentioned above create special difficulties in the asymptotic investigation of BVPs in thick junctions.

Literature review

As far as we know, the first works in this direction were papers by E.Ya. Khruslov [60], V.P. Kotlyarov and E.Ya. Khruslov [64], V.O. Marchenko and E.Ya. Khruslov [73, Sect. 5], in which convergence theorems were proved for the Green functions of Neumann problems for the Helmholtz equation in unbounded thick junctions of type 3:2:1. This was done either under assumptions of the convergence of certain components of the problem, or in case when the junction's body is a half-space. It should be noted that thick junctions considering in those papers have no periodicity in the distribution of the thin domains forming the junction.

Spectral problems

The limit equations describing acoustic vibrations in a porous medium, made periodically by narrow parallel channels or by thin parallel sheets in a solid body (the thick junctions of the type 3:2:1 or 3:2:2, respectively), were obtained by using the homogenization technique (two-scale asymptotic expansion method) in [8, 41, 130]. In these papers, the authors have established some new qualitative properties of the homogenized equations: the corresponding Helmholtz equation is no longer elliptic; the operator which corresponds to the spectral limit problem is non-compact and, as a result, "... *an exhaustive study of the spectral properties will not be done*" (see [41, Sect. 3, p. 163]). It should be noted that convergence theorems and asymptotic estimates were not proved in these papers. The same results were obtained in [8] for problems in a thick junction of type 3:2:2.

Independently of the publications mentioned above, in the early 90s the complete asymptotic analysis of the Neumann spectral problems for the Laplace operator in thick junctions of types 2:1:1 in $\mathbb{R}^2$ and 2:1:1, 3:1:1 in $\mathbb{R}^3$ was carried out in [104–107]. Namely, the corresponding homogenized spectral problems were found, their spectrum structures were determined, with the help of the method of matching asymptotic expansions the asymptotic approximations for eigenfunctions and also for the eigenvalues were constructed, and the corresponding asymptotic error estimates were proved.

One of the main and unexpected results of these problems is as follows. The spectrum of the original problem is split into countable family of series with different limits that constitute the spectrum of the corresponding homogenized problem. So, for the Neumann spectral problem in a thick junction type 2:1:1 (see [104]) the limits of the eigenvalues from each series form the sequence of the discrete eigenvalues of the homogenized problem from a finite interval (these intervals are mutually disjoint) and the left ends of these intervals belong to the essential spectrum of the homogenized problem. Moreover, there are a countable set of gaps in the spectrum (*a gap* is a bounded open interval having an empty intersection with the spectrum, but its ends belong to it). Such complex asymptotic behavior of the eigenvalues substantiates mathematically the well-known "loss reduction" phenomenon in comb-like waveguides [56] (for more detail see [107, 108]).

The question of the existence of spectral gaps has been actively investigated in last time since it is very important for the description of wave propagations in different mediums (see [53] for a lot of examples and references on this topic). In the case when the thin rods forming a thick junction have various length, the structure of the essential spectrum is more complex [82, 84] and choosing appropriately different lengths of the thin rods one can build a thick junction with the given number of gaps in the spectrum.

It is turned out that type of a thick junction determines not only the structure of the boundary layer in the junction zone, but also the principal properties of the homogenized problem, in particular, the structure of its spectrum. So, if the limit dimensions of the junction's body and the contact zone differ by 1 (i.e., $m - k = 1$), then it is possible to reduce the homogenized spectral problem to a spectral problem for some discontinuous self-adjoint operator-function. The spectrum of such operator-functions was studied in [51, 75, 108].

If we consider some boundary-value problem in a thick junction of the type $m : k : d$, where $m - k \neq 1$, then any transmission conditions are not reasonable in the joint zone for the corresponding homogenized problem in the Sobolev space $W^{1,p}$. For instance, if the junction in question is of type 3:1:1, then the junction zone becomes a curve Γ on the three-dimensional junction's body in the limit. In [105, 109], the spectral Neumann problem in a thick junction of type 3:1:1 was examined with simultaneous perturbation of the rigidity and density of the thin cylinders. Because of this some asymptotic "stratification" of the spectrum in question appears. The spectrum of the corresponding homogenized problem turns out to be related to the spectrum of a certain integral operator on Γ. Without the perturbation of the rigidity and density of the thin cylinders, the spectral Neumann problem was considered in [78, 88]. For this problem, we have the loss of self-adjointness in the limit in general. In fact, three spectral problems form the asymptotics for the eigenvalues and eigenfunctions of this problem, namely, the Neumann spectral problem in the junction's body; a spectral problem in a plane domain that is filled up by the thin cylinders in the limit (each eigenvalue of this problem has infinite multiplicity); and a spectral problem for a singular integral operator given on Γ.

Dirichlet spectral problems with purely density perturbations $\varepsilon^{-\alpha}$ on the rods of a thick junction of type 2:1:1 were studied in [77, 80, 86]. Depending on the parameter α ($\alpha < 2$, $\alpha = 2$, $\alpha > 2$), three qualitatively different cases in the asymptotic behavior of the eigenvalues and eigenfunctions were discovered. There are three kinds of vibrations, which are present in each of these cases: vibrations, whose energy is concentrated in the junction's body; vibrations, whose energy is concentrated on the thin rods; and vibrations (pseudo-vibrations) having quickly oscillating character, in which each thin rod can have its own frequency. The frequency range, where pseudo-vibrations can be present, is indicated using special characteristic

$$\mathscr{T} := \sup_{n \in \mathbb{N}} \limsup_{\varepsilon \to 0} \lambda_n(\varepsilon),$$

called the threshold of low eigenvalues. The threshold of low eigenvalues is equal to $+\infty$ for majority spectral problems of the perturbation theory including problems with concentrated masses in the cases $\alpha < 2$ and $\alpha = 2$; if $\alpha > 2$, then $\mathscr{T} = 0$. For spectral problems in thick junctions, the threshold $\mathscr{T}$ can be equal to some positive number [104, 106, 107].

It is known that the asymptotic behavior of the spectrum of spectral boundary-value problems in perturbed domains essentially depends on boundary conditions [7, 57, 76, 124]. Therefore, papers [81, 82, 84, 85] are devoted to the studying influence of different boundary conditions (Neumann, Robin, and Steklov) on the asymptotic behavior of eigenvalues of spectral problems in thick junctions of types 2:1:1, 3:2:1, and 3:2:2.

Boundary-value problems

The first asymptotic results for the Neumann Laplacian in thick junctions in $\mathbb{R}^N$ of the type $N : N - 1 : 1$ were obtained in the thesis by R. Brizzi and J.P. Chalot in 1978. After 19 years, these results were published in [15]. There it was proved that if the boundaries of thin cylinders forming a thick junction of type 3:2:1 are rectilinear, then the solution extended by zero converges to the corresponding homogenized problem. This is explained by the fact that this extension preserves the weak derivative with respect to x_3 due to the rectilinearity of the boundaries of the cylinders along the Ox_3-axis. This approach was used for homogeneous Neumann problems in [9, 10, 14, 24, 117] for monotone differential operators (in particular for the p-Laplacian) in thick junctions of the type $N : N - 1 : 1$ and $N - 1 : N - 1 : 1$, for the Ginzburg–Landau equation in a thick plane junction of type 2:2:1 [47], and for the wave equation [34].

Also in [15], the homogeneous Neumann problem was considered in a bounded plane domain whose boundary is waved by the function $x_2 = h(x_1/\varepsilon)$, where h must be a continuously differentiable periodic function and the reciprocal functions of h on some intervals have to be existed to construct special extension operator. However, this extension does not preserve the space class of the solution (only in $H^1_{loc}(\Omega_1^+)$, where $\Omega_1^+ \subset \mathbb{R}^2$ is a domain that is filled up by the oscillating boundary in the limit) and it was constructed under the assumption that the right-hand side $f \in H^1(\Omega_1)$. Using this extension and a special transformation of a surface integral over the oscillating boundary, the nonhomogeneous Neumann problem

$$-\Delta u_\varepsilon + u_\varepsilon = f \ \text{ in } \Omega_\varepsilon, \ \ \partial_{\nu_\varepsilon} u_\varepsilon = \varepsilon^\lambda \ \text{ on } \partial\Omega_\varepsilon,$$

was studied in [43]. If $\lambda > 1$, the same homogenized problem was obtained as in the corresponding homogeneous Neumann problem in [15]. If $\lambda = 1$, an additional term appears in the right-hand side of the homogenized problem. And if $\lambda \in [0, 1)$, the weak solution satisfies inequalities

$$\exists c_1, c_2 \in (0, +\infty) \ \forall \varepsilon : \ \frac{c_1}{\varepsilon^{1-\lambda}} \leq \|u_\varepsilon\|_{H^1(\Omega_\varepsilon)} \leq \frac{c_2}{\varepsilon^{1-\lambda}}.$$

From these results, we again see that the asymptotic behavior of the solution essentially depends on the conditions given on the oscillating part of the boundary.

The scheme of the construction of the extension operator proposed in [15] is not applicable for thick junctions with thin rectilinear cylinders ("pillar type" domains) since the geometry of the oscillating boundary cannot be described by a function. In [79], a new scheme of the construction of extension operators $\{ E : H^1(\Omega_\varepsilon) \mapsto H^1(\Omega)\}_{\varepsilon>0}$ was developed for solutions of BVPs. The extension construction for the solution to the homogeneous Neumann problem for the Poisson equation in a thick junction of the type 3:2:1 needs weaker conditions for the right-hand side. In addition, this extension preserves the space class of the solution and the scheme is most general in conception and applicable for solutions of BVPs in thick junctions with more complex structure [29, 31, 32].

Moreover, with the help of the method of matching asymptotic expansions, the approximation for the solution is constructed and the corresponding asymptotic error estimate in the Sobolev space $H^1(\Omega_\varepsilon)$ is proved in [79]. Similar results were obtained in [83] for parabolic problems with the linear Robin condition $\partial_{\nu_\varepsilon} u_\varepsilon + \varepsilon u_\varepsilon = 0$ given on the lateral surfaces of the thin cylinders, in [26] with the nonhomogeneous Robin condition

$$\partial_{\nu_\varepsilon} u_\varepsilon + \varepsilon^\alpha u_\varepsilon = \varepsilon^\beta g_\varepsilon(x, t) \ \ (\alpha, \beta \geq 1)$$

given on the lateral boundaries of the thin rings of a thick junction of type 3:2:2, and in [90] for the Poisson equation with the nonlinear Robin boundary condition

$$\partial_\nu u_\varepsilon + \varepsilon \kappa(u_\varepsilon) = 0$$

given on the lateral surfaces of the thin curvilinear cylinders of a thick junction of type 3:2:1. In those papers, it was showed how Robin boundary conditions are transformed (as $\varepsilon \to 0$) in the corresponding terms of the homogenized problem. In addition, these researches and ones made for spectral problems show that the method of matching asymptotic expansions introduced by Il'in [54, 55] is a very effective tool in the study of BVPs in thick junctions.

Obviously, if the homogeneous Dirichlet condition given on the oscillating boundary of a thick junction and the L^p-norm of the right-hand side of a boundary-value problem is bounded with respect to the parameter ε, then the zero extension of the solution converges to zero in $L^2(\Omega^+)$ (Ω^+ is a domain that is filled up by the oscillating boundary in the limit). The asymptotic approximation for the solution to the homogeneous Dirichlet problem for the Poisson equation was constructed and the corresponding H^1-asymptotic estimate outside ε-layer of the joint zone of a plane thick junction of type 2:1:1 was proved in [4]. It should be noted that the H^1-asymptotic estimates in the whole junction for the eigenfunctions of the Dirichlet Laplacian were derived in [80]. Using a boundary layer corrector, a first-order asymptotic approximation and H^1-asymptotic estimate (now in the whole junction) for the viscous incompressible flow governed by the stationary Stokes equations in a thick junction of type 3:2:1 with the homogeneous Dirichlet conditions on the oscillating boundary were obtained in [5, 6].

The asymptotic behavior of the transverse displacement of a Kirchhoff–Love plate in the form of a thick junction of type 1:1:1 was investigated in [13]. The main difficulty in analyzing this problem is twofold and it is essentially due to the fourth order of the Kirchhoff–Love model. Deriving a priori estimates on the displacement for this operator and for the oscillating domain after rescaling is more intricate than for second-order problems. Second, as far as the homogenization process is concerned, the use of the method of oscillating test functions, introduced by Tartar in 1977 and partially written in [118], is also more complicated than for second-order problems since here we have to take into account also the oscillations of the second derivatives of the solution.

Using the two-scale convergence technique, the homogenized results are obtained for a two-dimensional electrostatic model in thick junctions of type 1:1:1 that simulate the rotor and the stator of this model [48].

In [11, 12], the new "unfolding technique" introduced in [21] is adapted to prove convergence theorems for solutions of the elasticity system in thick junctions of types 3:2:1 and 2:2:1, respectively, with the help of a decomposition of the displacement field in the thin rods. With the help of this approach the convergence results are deduced for the elliptic problems in thick junctions with different boundary conditions (homogeneous Neumann, nonhomogeneous Neumann, Robin) on the oscillating boundary (see [2, 72]).

The asymptotic behavior of second-order elliptic and parabolic Neumann problems in thick junctions Ω_ε of the type $N : N - 1 : 1$ under the following assumptions for the right-hand side:

$$f_\varepsilon \in L^2(\Omega), \quad \exists c > 0 \; \forall \varepsilon : \int_{\Omega_\varepsilon} (1 + |f_\varepsilon|) \ln(1 + |f_\varepsilon|) \, dx \leq c$$

was studied in [44, 45]. If the the right-hand side $f_\varepsilon \in L^1(\Omega)$ and if no periodicity is assumed on the distribution of the cylindrical vertical teeth of a thick junction, the asymptotic behavior of the solution to an elliptic homogeneous Neumann problem was investigated in [46].

The homogenization of optimal control problems for elliptic, parabolic, and wave equations with homogeneous Neumann boundary conditions in thick junctions of the type $N : N - 1 : 1$ with different cost functionals was made in [28, 35, 61, 62]. With the help of the unfolding method, the homogenization of optimal control problems in thick junctions with the homogeneous Neumann conditions on the oscillating boundary were made in [119, 120, 126].

The Signorini conditions most closely correspond to the modeling of different processes in domains with the complex boundary structure since in fact it is impossible to control boundary conditions on a such boundary. The classical Signorini conditions contain two alternative linear boundary conditions (Dirichlet or Neumann) and it is unknown which of these conditions is satisfied for each point of the boundary. Homogenization of both elliptic and parabolic Signorini boundary-value problems in thick junctions of types 2:1:1 and 3:2:1 was made in [58, 101, 103], where it was showed that the Signorini conditions are transformed (as $\varepsilon \to 0$) in differential inequalities in the region that is filled up by the thin domains.

In [102], the following nonlinear boundary conditions of the Signorini type:

$$u_\varepsilon \leq g, \quad \partial_\nu u_\varepsilon + \varepsilon^\alpha h(u_\varepsilon) \leq 0, \quad (u_\varepsilon - g)\,(\partial_\nu u_\varepsilon + \varepsilon^\alpha h(u_\varepsilon)) = 0 \qquad (1.1)$$

on the lateral surfaces of the thin cylinders of a thick junction were considered. Here the parameter $\alpha \in \mathbb{R}$, the function h describing the nonlinearity belongs to $W_{loc}^{1,\infty}(\mathbb{R})$ and there exist positive constants c_1 and c_2 such that $c_1 \leq h'(s) \leq c_2$ for a.e. $s \in \mathbb{R}$, and $h(0) = 0$ (so-called the zero-absorption condition) if $\alpha < 1$. We see that the passage to the limit is accompanied by special intensity factor ε^α in the boundary conditions. Two qualitatively different cases in the asymptotic behavior of the solution are established depending on the value of the parameter α, namely $\alpha \geq 1$ and $\alpha < 1$. For each case, the convergence theorem is proved.

In addition, the convergence of the energy integrals for elliptic Signorini problems were obtained in those two cases. It was pointed out in [134] that in fact there is only one natural definition of homogenization for functionals that are defined on reflexive spaces and grow faster than the norm—the definition via the convergence of energies. Therefore, the proof of the convergence of the energy integrals is a very important result that makes it possible to study optimal control problems in thick junctions by using the direct method of the calculus of variations.

A nonlinear monotone problem with nonlinear Signorini boundary conditions

$$\begin{cases} u_\varepsilon(x) \leq g(x), \quad \mathbf{A}\,(x, Du_\varepsilon(x)) \cdot v_\varepsilon(x) + \varepsilon^\lambda h(x, u_\varepsilon(x)) \leq 0, \\ (u_\varepsilon(x) - g(x))\,\big(\mathbf{A}(x, Du_\varepsilon(x)) \cdot v_\varepsilon(x) + \varepsilon^\lambda h(x, u_\varepsilon)\big) = 0 \end{cases}$$

on the lateral surfaces of the thin rectilinear cylinders of a thick junction of the type $N : N - 1 : 1$ is considered in [49, 50]. The main assumption for the monotone function h with respect the second variable is as follows:

$$\exists \eta > 0, \ \eta_1 \in W^{1, \frac{p}{p-1}}(\Omega^+) \ : \ |h(x, s)| \le \eta |s|^{p-1} + \eta_1(x), \ \text{a.e.} \ x \in \Omega^+, \ \forall s \in \mathbb{R}.$$

If $\lambda \ge 1$ the corresponding homogenized problem is nonstandard since it is given by a variational inequality coupled to an algebraic system in the region filled up by the thin cylinders [49]. For $\lambda \in (0, 1)$, the additional assumption

$$\exists \delta > 0 : \ \delta |s|^p \le h(x, t) s, \quad \text{a.e.} \ x \in \Omega^+, \ \forall s \in \mathbb{R}.$$

is needed to prove the convergence theorem [50]; the limit problem consists of a boundary-value problem in the junction's body and also an algebraic system in the region filled up by the thin cylinders. In the case $\lambda < 0$, the convergence results are obtained under more strong assumption on the vector-function $\mathbf{A}$ (see [50]).

Thick multilevel junctions

Successful applications of thick-junction constructions have stimulated active studying of BVPs in thick junctions with more complex structures. *A thick multilevel junction* is a thick junction in which the thin domains are divided into a finite number of levels depending on their geometric structure and boundary conditions imposed on their boundaries. Besides, the thin domains from each level ε-periodically alternate along the joint zone.

For the first time, a boundary-value problem in a plane thick multilevel junction of type 2:1:1 was considered in [87], where the asymptotic behavior of eigenvalues and eigenfunctions was analyzed (see also [89, 91]). In [33], with the help of the special multilevel extension operator the convergence theorem was proved for the solution to the Poisson equation in a plane two-level junction with homogeneous Robin boundary conditions at the boundaries of the thin rods. Using the method of matched asymptotic expansions, the first terms of the asymptotics were constructed for solutions both to elliptic and to parabolic BVPs with different varying types of boundary conditions in thick multilevel junctions of types 2:1:1 and 3:2:1 in [27, 36, 39, 114, 115], and in addition, the corresponding asymptotic estimates in the Sobolev space $H^1(\Omega_\varepsilon)$ (in $L^2(0, T; H^1(\Omega_\varepsilon))$) for parabolic problems) were proved there. Homogenization of a semilinear Signorini problem in a thick multilevel junction of type 3:2:1, was made in [102].

Using the Buttazzo-Dal Maso abstract scheme for variational convergence of constrained minimization problems, the asymptotic analysis of optimal control problems (linear and quasilinear) with inhomogeneous perturbed Robin boundary conditions in multilevel junctions of types 2:1:1 and 3:2:1 was made in [37, 38]. Qualitatively different cases in the asymptotic behavior of the solutions were discovered and an interesting application for an optimal control problem involving a thick one-level junction with cascade controls is presented in [38].

From the results obtained in these papers, it follows that physical processes in thick multilevel junctions behave as a "many-phase system" in the region which is filled up by the thin domains from each level in the limit.

Thick cascade junctions

By contrast with the papers that studied BVPs in thick multilevel junctions with the thin offshoots of finite length, new effects emerge for BVPs in *thick cascade junctions* (it is a thick multilevel junction in which the thin domains of the first level have the fixed length of order $\mathscr{O}(1)$ and the thin domains of the second one vanish as $\varepsilon \to 0$). For such problems new kind of inhomogeneous conjugation conditions appeared in the corresponding homogenized problem at the joint zone, which take into account the geometry of the thin domains of the vanishing level and boundary conditions given on their boundaries (see [95–97], where the homogenization theorems and the convergence of the energy integrals were proved for BVPs with different boundary conditions in thick cascade junctions of types 2:1:1 and 3:2:1). In [100], similar results were obtained in a plane thick cascade junction in which the thin rectangles from the second level have the length of order ε^{α}, $(0 < \alpha < 1)$, i.e., the length of a rectangle from the second level is small, but considerably larger than the alternation period ε.

BVPs in new kind of thick cascade junctions were studied in [16, 17]. Such junctions consist of a body and a large number of thin rectangles (cylinders) joining to the body through the thin random transmission zone with rapidly oscillating boundary. Assuming the inhomogeneous Robin boundary conditions with perturbed coefficients to be set on the boundaries of the thin cylinders and with random perturbed coefficients on the boundary of the transmission zone, the homogenization theorems and the convergence of the respective energy integrals were proved there. It is shown that there are three qualitatively different cases in the asymptotic behavior of the solutions and the energy integrals.

A rich collection of new results on asymptotic analysis of spectral problems in thick cascade junctions with concentrated masses on the thin rods of the second vanishing level (the density is order $\mathscr{O}(\varepsilon^{-\alpha})$ $(\alpha > 0)$ on the thin rods from the second level and $\mathscr{O}(1)$ outside of them) are obtained in [18–20, 98, 99]. In contrast with [86], five qualitatively different cases in the asymptotic behavior of eigen-magnitudes (as $\varepsilon \to 0$) are discovered, namely the case of "light" concentrated masses ($\alpha \in (0, 1)$), "middle" concentrated masses ($\alpha = 1$), and "heavy" concentrated masses ($\alpha \in (1, +\infty)$) that is divided into "slightly heavy" concentrated masses ($\alpha \in (1, 2)$), "intermediate heavy" concentrated masses ($\alpha = 2$), and "very heavy" concentrated masses ($\alpha > 2$).

If $\alpha \in (0, 1)$, then the spectrum of the homogenized problem coincides with the spectrum of the problem in the domain without concentrated masses (see, for instance, [81, 106] mentioned above). The concentrated masses bring the influence only from the second terms of the asymptotic expansions for the eigen-elements. Concentrated masses first manifest themselves in the homogenized spectral problem for $\alpha = 1$. Namely, we observe an additional summand with a spectral parameter in the jump of derivatives at the joint zone. If $\alpha > 1$, then the concentrated masses begin to play a principal role in the asymptotic behavior of the eigenvalues and eigenfunctions. A fundamental difference between this case and the previous ones is that all eigenvalues decay at the rate of $\varepsilon^{\alpha-1}$. The corresponding eigen-vibrations have a new type of the skin effect (surface waves appearing in the solid) that was called *spatial-skin effect*. It means that the vibrations of the thin long rods of the first level repeat the shape of the vibrations of the joint zone. For $\alpha \geq 2$, the spatial-skin effect for the corresponding eigenfunctions is strengthened; in addition, the geometry of the junction's body and thin rectangles of the first level has a smaller influence on the asymptotic behavior of the eigenvalues. Therefore, the case of "heavy" concentrated masses is divided into three ones.

For all these cases, the Hausdorff, low- and high-frequency convergences of the spectrum to the spectrum of the corresponding homogenized problem as $\varepsilon \to 0$ were proved, the leading terms of the asymptotics both for the eigenfunctions and eigenvalues were constructed and justified.

Thick junctions with the branched fractal structure

In [92, 93], the first author have begun to study asymptotic properties of solutions to semilinear parabolic initial-boundary-value problems in *thick fractal junctions* (or *thick junctions with branched structure*). A thick fractal junction is the union of the junction's body and a lot of thin trees situated ε-periodically along the joint zone. The trees have finite number of branching levels. In particular, the following nonlinear Robin boundary condition $\partial_\nu v_\varepsilon + \varepsilon^{\alpha_i} \mu(t, x_2, v_\varepsilon) = \varepsilon^\beta g_\varepsilon$ on the branch boundaries of the ith branching layer was considered in [92]; $\{\alpha_i\}$ and β are real parameters. The asymptotic analysis of such problems was made as $\varepsilon \to 0$, i.e., when the number of the thin trees infinitely increases and their thickness vanishes. Namely, for each such problem a corresponding homogenized problem was found and the existence and uniqueness of its solution in an anisotropic Sobolev space of multi-sheeted functions were proved. Also the asymptotic approximation for the solution was constructed and the corresponding asymptotic estimate in the space $C([0, T]; L^2(\Omega_\varepsilon)) \cap L^2(0, T; H^1(\Omega_\varepsilon))$ was proved.

Using the unfolding operator method, convergence results for an optimal control problem involving a linear parabolic problem with the homogeneous Neumann condition in a thick junction with branched structure are obtained in [1]; and more general results for an optimal control problem involving a linear elliptic problem with the controls on the oscillating part are obtained in [3].

Overview of the book

In spite of significant achievements in studying BVPs in thick junctions, problems in thick multilevel junctions of type 3:2:2 are not studied properly. Therefore, this book is devoted to the study of the asymptotic behavior of solutions to BVPs in such thick multilevel junctions. The results presented here have already been published in our articles [110–113, 127–129], but we give here a complete and unified presentation.

The Dirichlet, Neumann, and Robin (Fourier) boundary conditions are considered on the boundaries of the thin annular discs that strung on a fixed cylinder that is the junction's body of the thick multilevel junction Ω_ε. Additionally Robin boundary conditions depend on perturbation parameters. We study the influence of various combinations of such boundary conditions given on the surfaces of the thin discs from different levels, as well as the influence of the perturbation parameters on the asymptotic behavior of solutions of BVPs in Ω_ε.

In Chap. 2, convergence theorems are proved for solutions to linear elliptic BVPs in Ω_ε. Here we consider two BVPs, the first one is with various alternating perturbed Robin boundary conditions, and the second one is with alternating Neumann and Dirichlet boundary conditions on the surfaces of the thin discs. The convergence of the energy integrals for each problem is also proved. As noted above, this is a very useful result that gives possibility directly obtain results for optimal control problems involving thick multilevel junctions of type 3:2:2.

In Chap. 3, we consider the case when the thin domains of a thick junction can have sharp edges, i.e., their thickness tends to zero polynomially with the exponent $1 + \gamma$ while approaching the edges (here $\gamma > -1$ is a parameter). Three qualitatively different cases in the asymptotic behavior of the solution are discovered depending on the edge form, namely rounded edges $\gamma \in (-1, 0)$, linear wedges $\gamma = 0$, and very sharp edges $\gamma > 0$ (in this case the boundary is not Lipshitz). Nonstandard and new techniques are proposed to get the corresponding homogenized problems (untypical in the cases $\gamma = 0$ and $\gamma > 0$). The obtained results mathematically justify an interesting physical effect for heat radiators (see conclusions to this chapter).

It should be noted here that thick junctions are domains with Lipshitz boundary in all papers mentioned in this review (in some articles the oscillating boundaries must be smooth).

In Chap. 4, the method proposed in Chap. 2 is broadened for semilinear parabolic BVPs in Ω_ε. Here we show how to apply the Minty–Browder method to homogenize nonlinear Robin conditions that have special intensity factor ε^α, where the parameter $\alpha \in \mathbb{R}$ and its significant impacts on the asymptotic behavior of the solutions.

In Chap. 5, approximation techniques are demonstrated for semilinear elliptic and parabolic problems in Ω_ε with various alternating Robin boundary conditions. With the help of special junction-layer solutions, whose behavior is determined by type 3:2:2, and the method of matched asymptotic expansions, approximation functions are constructed for the solutions and the corresponding asymptotic estimates in Sobolev spaces are proved. These estimates show the influence of the given data and parameters on the asymptotic behavior of the solutions.

It should be noted that the results obtained in each chapter are discussed, including their physical interpretations, in Conclusions to each chapter.

Chapter 2
Homogenization of Linear Elliptic Problems

2.1 Statement of Problems

Let r_0, r_1, r_2, l_1, l_2 be fixed positive numbers such that $r_0 < r_1 \leq r_2$ and $l_1 < l_2 < 1$; and let $h_i : [r_0, r_i] \mapsto (0, 1)$ be piecewise smooth functions satisfying the following conditions:

$$0 < l_i - \frac{h_i(s)}{2}, \quad l_i + \frac{h_i(s)}{2} < 1 \quad \forall s \in [r_0, r_i], \ i = 1, 2,$$

$$l_1 + \frac{h_1(s)}{2} < l_2 - \frac{h_2(s)}{2} \quad \forall s \in [r_0, r_1].$$

These inequalities imply that for all $s \in [r_0, r_i]$ the intervals

$$I_i(s) := \left(l_i - \frac{h_i(s)}{2}, \ l_i + \frac{h_i(s)}{2} \right), \quad i = 1, 2,$$

are contained inside $(0, 1)$ and $\overline{I_1(s)} \cap \overline{I_2(s)} = \emptyset$.

Let us introduce a cylindrical coordinate system (r, x_2, θ) in $\mathbb{R}^3$:

$$x = (r \cos \theta, \ x_2, \ r \sin \theta), \quad r = \sqrt{x_1^2 + x_3^2}, \ x_2 \in \mathbb{R}, \ \theta \in [0, 2\pi).$$

Remark 2.1 Hereinafter passing to the cylindrical coordinate system, we will keep the same notations for functions, i.e., $\varphi(x) = \varphi(r, x_2, \theta)$ if (r, x_2, θ) are the corresponding cylindrical coordinates of the point $x = (x_1, x_2, x_3) \in \mathbb{R}^3$.

Let N be a large positive integer and l be a fixed positive number; then $\varepsilon = l/N$ is a small parameter. We denote by $I_\varepsilon^{(i)}(j, a)$ the interval

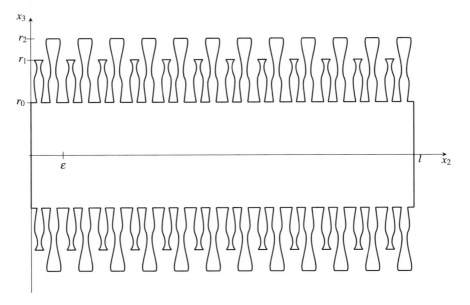

Fig. 2.1 A cross section of the thick junction Ω_ε of type 3:2:2 ($N = 12$)

$$\left(\varepsilon(j + l_i - \tfrac{a}{2}), \ \varepsilon(j + l_i + \tfrac{a}{2})\right)$$

of the length εa, where $i \in \{1, 2\}$, $a \in (0, 1)$, and $j \in \{0, 1, \ldots, N - 1\}$. Obviously, for every $i \in \{1, 2\}$ and $r \in [r_0, r_i]$ the intervals $\{I_\varepsilon^{(i)}(j, h_i(r))\}_{j=\overline{0, N-1}}$ are ε-periodically situated inside $(0, l)$.

A thick multilevel junction Ω_ε (see Fig. 2.1) consists of a cylinder

$$\Omega_0 = \{x \in \mathbb{R}^3 : 0 < x_2 < l, \ r < r_0\}$$

and a large number of thin annular discs

$$\Omega_\varepsilon^{(1)} = \bigcup_{j=0}^{N-1} \Omega_\varepsilon^{(1)}(j), \quad \Omega_\varepsilon^{(2)} = \bigcup_{j=0}^{N-1} \Omega_\varepsilon^{(2)}(j),$$

i.e., $\Omega_\varepsilon = \Omega_0 \cup \Omega_\varepsilon^{(1)} \cup \Omega_\varepsilon^{(2)}$. Here the jth thin disc from the ith level is defined as follows:

$$\Omega_\varepsilon^{(i)}(j) := \{x \in \mathbb{R}^3 : x_2 \in I_\varepsilon^{(i)}(j, h_i(r)), \ r_0 \le r < r_i\},$$

where $i = 1, 2; \ j = \overline{0, N - 1}$.

Remark 2.2 Here the junction's body is the 3-dimensional cylinder Ω_0, the joint zone is its lateral surface, and each of the attached thin annular discs is shrunk into an annulus as $\varepsilon \to 0$. Thus, the type of the thick junction Ω_ε is 3:2:2.

The parameter ε characterizes distance between neighboring thin discs, their quantity, and their thickness. The number of the discs equals $2N$, they are divided into two levels $\Omega_\varepsilon^{(1)}$ and $\Omega_\varepsilon^{(2)}$, and the discs from each level ε-periodically alternate.

In what follows we will often use the following notations:

- $\partial\Omega_\varepsilon^{(i)}(j) \cap \{r_0 < r < r_i\}$ is the union of two lateral surfaces of the disc $\Omega_\varepsilon^{(i)}(j)$;
- $\partial\Omega_\varepsilon^{(i)}(j) \cap \{r = r_i\}$ is the outer edge of $\Omega_\varepsilon^{(i)}(j)$;
- $\Omega_\varepsilon^{(i)}(j) \cap \{r = r_0\}$ is the inner edge of $\Omega_\varepsilon^{(i)}(j)$;
- $\partial\Omega_\varepsilon^{(i)}(j) \cap \{r > r_0\}$ is the part of the boundary of $\Omega_\varepsilon^{(i)}(j)$ except of the inner edge;
- $\partial\Omega_\varepsilon \cap \{r = r_0\}$ is the part of the lateral surface of the cylinder Ω_0 situated between the inner edges of the thin discs;
- $\partial\Omega_0 \cap \{r < r_0\}$ are the bases of the cylinder Ω_0;
- $\Omega' = \partial\Omega_0 \cap \{r = r_0\}$ is the joint zone of the thick multilevel junction Ω_ε.

We consider two linear elliptic BVPs in Ω_ε. The first one is

$$
\begin{cases}
-\Delta_x u_\varepsilon(x) = f_\varepsilon(x), & x \in \Omega_\varepsilon, \\
\partial_\nu u_\varepsilon(x) + \varepsilon^\alpha k_1 u_\varepsilon(x) = \varepsilon^\beta g_\varepsilon(x), & x \in \partial\Omega_\varepsilon^{(1)} \cap \{r > r_0\}, \\
\partial_\nu u_\varepsilon(x) + \varepsilon k_2 u_\varepsilon(x) = \varepsilon^\beta g_\varepsilon(x), & x \in \partial\Omega_\varepsilon^{(2)} \cap \{r_0 < r < r_2\}, \\
\partial_\nu u_\varepsilon(x) + k_2 u_\varepsilon(x) = 0, & x \in \partial\Omega_\varepsilon^{(2)} \cap \{r = r_2\}, \\
\partial_\nu u_\varepsilon(x) = 0, & x \in \partial\Omega_\varepsilon \cap \{r = r_0\}, \\
\partial_\nu u_\varepsilon(x) = q_\varepsilon(x), & x \in \partial\Omega_0 \cap \{r < r_0\},
\end{cases}
\tag{2.1}
$$

with alternating singularly perturbed Robin boundary conditions on the surfaces of the thin discs from both levels; the second one is

$$
\begin{cases}
-\Delta_x v_\varepsilon(x) = f_\varepsilon(x), & x \in \Omega_\varepsilon, \\
\partial_\nu v_\varepsilon(x) = \varepsilon g_\varepsilon(x), & x \in \partial\Omega_\varepsilon^{(1)} \cap \{r > r_0\}, \\
v_\varepsilon(x) = 0, & x \in \partial\Omega_\varepsilon^{(2)} \cap \{r > r_0\}, \\
\partial_\nu v_\varepsilon(x) = 0, & x \in \partial\Omega_\varepsilon \cap \{r = r_0\}, \\
\partial_\nu v_\varepsilon(x) = q_\varepsilon(x), & x \in \partial\Omega_0 \cap \{r < r_0\}.
\end{cases}
\tag{2.2}
$$

with alternating inhomogeneous Neumann and homogeneous Dirichlet boundary conditions Here $\partial_\nu = \partial/\partial\nu$ is the outward normal derivative; $k_1 \geq 0$, $k_2 > 0$; and parameters $\alpha \geq 1$, $\beta \geq 1$.

Remark 2.3 (Comments to the statement) In a typical interpretation, solutions to BVPs represent the density of some quantity (chemical concentration, temperature, electronic potential) at equilibrium within a domain, where a boundary-value problem is considered. In many physical problems, the flow across the boundary surface is proportional to the difference between the surrounding density g and the density u just inside of the domain. As a result, we get the linear Robin boundary condition $\partial_\nu u + k_0 u = g$ on the boundary (see [125] for more detail). Here, in the book, we will interpret solutions as temperature inside Ω_ε. Then k_0 is a positive heat transfer coefficient.

There are intensity factors in ε^α, ε, and ε^β in the boundary conditions of the problem (2.1). They usually appear after the nondimensionalization of mathematical models in thin domains (see e.g. [94]). As will be shown below, the asymptotic behavior of the solution depends essentially on the parameters α and β. Moreover, two cases are qualitatively different, namely when they are equal to 1 and when they greater than 1. Therefore, for simplicity, we take $\alpha = 1$ in the boundary condition on $\partial \Omega_\varepsilon^{(2)} \cap \{r_0 < r < r_2\}$ (it can also mean that discs from the second level have different physical properties). The case $\alpha < 1$ is considered in Chap. 4. If $k_1 = 0$, then we have Neumann boundary conditions in (2.1).

Clearly, the outer edges of the thin discs from the second level are more exposed to outer environment than the other parts of the boundaries of the thin discs from both levels (see Fig. 2.1). Thus, it is natural to impose distinct boundary conditions on those parts. In (2.1), we impose Robin boundary conditions on the edges of $\Omega_\varepsilon^{(2)}$ with the heat transfer coefficient k_2.

Alternation of the thin discs from these two levels leads to rapid change of boundary conditions along the joint zone Ω' in BVPs (2.1) and (2.2).

It is easy to see that the thin annular discs from the ith level fill up the set

$$\Omega^{(i)} = \{x \in \mathbb{R}^3 : 0 < x_2 < l,\ r_0 \leq r < r_i\} \quad \text{as } \varepsilon \to 0\ (i = 1,\ 2).$$

The aim of this chapter is to answer what will happen with BVPs (2.1) and (2.2) as $\varepsilon \to 0$ ($N \to +\infty$), i.e., when the number of the attached thin discs of each level infinitely increases and their thickness tends to zero. We should find the corresponding homogenized in each case, prove the convergence of solutions, and study the influence of the parameters α and β.

As for the given functions, we suppose that the following conditions take place:

- the functions f_ε, $f_0 \in L^2(\Omega_0 \cup \Omega^{(2)})$, and

$$f_\varepsilon \longrightarrow f_0 \quad \text{strongly in } L^2(\Omega_0 \cup \Omega^{(2)}) \text{ as } \varepsilon \to 0; \tag{2.3}$$

- the functions g_ε, $g_0 \in H^1(\Omega^{(2)})$, and

$$g_\varepsilon \xrightarrow{w} g_0 \quad \text{weakly in } H^1(\Omega^{(2)}) \text{ as } \varepsilon \to 0; \tag{2.4}$$

- the functions q_ε, $q_0 \in L^2(\partial \Omega_0 \cap \{r < r_0\})$, and

$$q_\varepsilon \xrightarrow{w} q_0 \quad \text{weakly in } L^2(\partial \Omega_0 \cap \{r < r_0\}) \text{ as } \varepsilon \to 0. \tag{2.5}$$

Definition 2.1 A function $u_\varepsilon \in H^1(\Omega_\varepsilon)$ is called a weak solution to problem (2.1) if the integral identity

$$(u_\varepsilon,\ \varphi)_{1,\varepsilon} = \mathsf{L}_{1,\varepsilon}(\varphi) \tag{2.6}$$

holds for all $\varphi \in H^1(\Omega_\varepsilon)$, where

$$(u_\varepsilon, \varphi)_{1,\varepsilon} = \int_{\Omega_\varepsilon} \nabla_x u_\varepsilon \cdot \nabla_x \varphi \, dx + \varepsilon^\alpha k_1 \int_{\partial\Omega_\varepsilon^{(1)} \cap \{r > r_0\}} u_\varepsilon \varphi \, d\sigma_x$$

$$+ \varepsilon k_2 \int_{\partial\Omega_\varepsilon^{(2)} \cap \{r_0 < r < r_2\}} u_\varepsilon \varphi \, d\sigma_x + k_2 \int_{\partial\Omega_\varepsilon^{(2)} \cap \{r = r_2\}} u_\varepsilon \varphi \, d\sigma_x$$

and

$$\mathsf{L}_{1,\varepsilon}(\varphi) = \int_{\Omega_\varepsilon} f_\varepsilon \varphi \, dx + \varepsilon^\beta \int_{\partial\Omega_\varepsilon^{(1)} \cap \{r > r_0\} \cup \partial\Omega_\varepsilon^{(2)} \cap \{r_0 < r < r_2\}} g_\varepsilon \varphi \, d\sigma_x + \int_{\partial\Omega_0 \cap \{r < r_0\}} q_\varepsilon \varphi \, d\tilde{x}.$$

Remark 2.4 Hereinafter $\tilde{x} = (x_1, x_3)$.

Consider a Sobolev space

$$H^1(\Omega_\varepsilon; \partial\Omega_\varepsilon^{(2)} \cap \{r > r_0\}) := \{\varphi \in H^1(\Omega_\varepsilon) : \varphi|_{\partial\Omega_\varepsilon^{(2)} \cap \{r > r_0\}} = 0\}.$$

Hereinafter $\varphi|_S$ denotes either the trace of a function φ on a surface $S \subset \mathbb{R}^3$ or the restriction of φ on a domain $S \subset \mathbb{R}^3$.

Definition 2.2 A function $v_\varepsilon \in H^1(\Omega_\varepsilon; \partial\Omega_\varepsilon^{(2)} \cap \{r > r_0\})$ is called a weak solution to problem (2.2) if it satisfies the identity

$$(v_\varepsilon, \varphi)_{2,\varepsilon} = \mathsf{L}_{2,\varepsilon}(\varphi) \tag{2.7}$$

for all functions $\varphi \in H^1(\Omega_\varepsilon; \partial\Omega_\varepsilon^{(2)} \cap \{r > r_0\})$, where

$$(v_\varepsilon, \varphi)_{2,\varepsilon} = \int_{\Omega_\varepsilon} \nabla_x v_\varepsilon \cdot \nabla_x \varphi \, dx,$$

$$\mathsf{L}_{2,\varepsilon}(\varphi) = \int_{\Omega_\varepsilon} f_\varepsilon \varphi \, dx + \varepsilon \int_{\partial\Omega_\varepsilon^{(1)} \cap \{r > r_0\}} g_\varepsilon \varphi \, d\sigma_x + \int_{\partial\Omega_0 \cap \{r < r_0\}} q_\varepsilon \varphi \, d\tilde{x}.$$

2.2 Auxiliary Statements and a Priori Estimates

2.2.1 Auxiliary Statements

It is easy to check that for a.e. $x \in \partial\Omega_\varepsilon^{(i)} \cap \{r_0 < r < r_i\}$ the outward unit normal to the lateral surface of the thin disc from the ith level is

$$\nu(x) = \left(N_\varepsilon^{(i)}(r)\right)^{-1}\left(-\frac{\varepsilon \partial_{x_1} h_i(r)}{2}, \pm 1, -\frac{\varepsilon \partial_{x_3} h_i(r)}{2}\right), \quad i = 1, 2, \tag{2.8}$$

where $N_\varepsilon^{(i)}(r) = \sqrt{1 + 4^{-1}\varepsilon^2 |h_i'(r)|^2}$, and "+" (or "−") corresponds to the right (or left) part of the lateral surface of the thin disc $\Omega_\varepsilon^{(i)}(j)$. Further we will be using the following integral identity proved in [32]:

$$\frac{\varepsilon}{2} \int_{\partial\Omega_\varepsilon^{(i)} \cap \{r_0 < r < r_i\}} \frac{h_i(r)\varphi}{N_\varepsilon^{(i)}(r)} \, d\sigma_x = \int_{\Omega_\varepsilon^{(i)}} \varphi \, dx - \varepsilon \int_{\Omega_\varepsilon^{(i)}} Y_i \left(\frac{x_2}{\varepsilon}\right) \partial_{x_2}\varphi \, dx \qquad (2.9)$$

for any $\varphi \in H^1(\Omega_\varepsilon^{(i)})$; here $Y_i(s) = -s + [s] + l_i$ and $[s]$ is the integer part of $s \in \mathbb{R}$. Identity (2.9) and the estimate $\max_\mathbb{R} |Y_i(s)| \leq 2$ provide the inequalities

$$\|\varphi\|_{L^2(\partial\Omega_\varepsilon^{(i)} \cap \{r_0 < r < r_i\})} \leq C_0 \varepsilon^{-\frac{1}{2}} \|\varphi\|_{H^1(\Omega_\varepsilon^{(i)})}, \qquad (2.10)$$

$$\int_{\Omega_\varepsilon^{(i)}} \varphi^2 \, dx \leq C_1 \left(\varepsilon^2 \int_{\Omega_\varepsilon^{(i)}} |\nabla_x\varphi|^2 \, dx + \varepsilon \int_{\partial\Omega_\varepsilon^{(i)} \cap \{r_0 < r < r_i\}} \varphi^2 \, d\sigma_x \right) \qquad (2.11)$$

for all $\varphi \in H^1(\Omega_\varepsilon^{(i)})$, $i = 1, 2$. Obviously, inequality (2.11) for functions $\varphi \in H^1(\Omega_\varepsilon, \partial\Omega_\varepsilon^{(2)} \cap \{r > r_0\})$ looks like

$$\int_{\Omega_\varepsilon^{(2)}} \varphi^2 \, dx \leq C_1 \varepsilon^2 \int_{\Omega_\varepsilon^{(2)}} |\nabla_x\varphi|^2 \, dx. \qquad (2.12)$$

Remark 2.5 Hereafter all constants c_i, C_i in inequalities are positive and independent of ε.

Lemma 2.1 *For any function $\varphi \in H^1(\Omega_\varepsilon^{(i)})$, the following inequalities take place:*

$$\|\varphi\|_{L^2(\partial\Omega_\varepsilon^{(i)} \cap \{r=r_i\})} \leq C_2 \|\varphi\|_{H^1(\Omega_\varepsilon^{(i)})}, \qquad (2.13)$$

$$\|\varphi\|_{L^2(\Omega_\varepsilon^{(i)} \cap \{r=r_0\})} \leq C_3 \|\varphi\|_{H^1(\Omega_\varepsilon^{(i)})}, \qquad (2.14)$$

$$\|\varphi\|_{L^2(\Omega_\varepsilon^{(i)})} \leq C_4 \big(\big\| |\nabla_x\varphi| \big\|_{L^2(\Omega_\varepsilon^{(i)})} + \|\varphi\|_{L^2(\Omega_\varepsilon^{(i)} \cap \{r=r_0\})} \big), \quad i = 1, 2. \qquad (2.15)$$

Proof We prove (2.13) by using the technique of [96, Lemma 2.1]. Fix $\delta > 0$ such that the interval $(\check{x}_\varepsilon, \hat{x}_\varepsilon)$ is contained inside $\Omega_\varepsilon^{(1)}(j)$ (see Fig. 2.2), where $\hat{x}_\varepsilon = (r_1, \varepsilon(j + l_1 - \frac{h_1(r_1)}{2}), \theta)$, $\check{x}_\varepsilon = (r_1 - \delta, \varepsilon(j + l_1 - \frac{h_1(r_1)}{4}), \theta)$. Obviously, δ depends only on the function $h_1(r)$; here $j \in \{0, \ldots, N - 1\}$, and $\theta \in [0, 2\pi)$.

For any function $\varphi \in C^\infty(\Omega_\varepsilon^{(1)})$, $r \in (r_1 - \delta, r_1)$, and $x_2 \in I_\varepsilon^{(1)}(j, h_1(r))$, we have

$$\varphi(r_1, x_2, \theta) = \int_{(r, (x_2+\varepsilon(j+l_1))/2, \theta)}^{(r_1, x_2, \theta)} \partial_s\varphi \, ds + \varphi(r, (x_2 + \varepsilon(j + l_1))/2, \theta),$$

where s is a natural parameter of the interval $\big((r, (x_2 + \varepsilon(j + l_1))/2, \theta), (r_1, x_2, \theta)\big)$. Squaring this equality and using the Cauchy–Schwartz inequality, we get

Fig. 2.2 A cross section of the thin disc $\Omega_\varepsilon^{(1)}(j)$

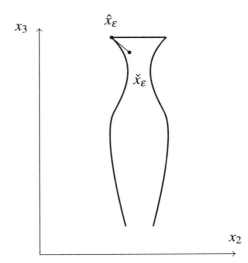

$$\varphi^2(r_1,\, x_2,\, \theta) \le c_0 \int_{(r,\, (x_2+\varepsilon(j+l_1))/2,\, \theta)}^{(r_1,\, x_2,\, \theta)} |\nabla_x \varphi|^2 \, ds + 2\varphi^2(r,\, (x_2 + \varepsilon(j+l_1))/2,\, \theta).$$

Now integrating this inequality in $r \in (r_1 - \delta,\, r_1)$, $x_2 \in I_\varepsilon^{(1)}(j,\, h_1(r))$, and $\theta \in [0,\, 2\pi)$, then summing over $j \in \{0, \ldots, N-1\}$, we arrive (2.13). By the closure, it remains valid for all $\varphi \in H^1(\Omega_\varepsilon^{(1)})$.

Similarly, we prove (2.13) for the thin discs of the second level and the other estimates (2.14) and (2.15). The Lemma is proved.

If $\mathscr{S}$ is a Lebesgue-measurable subset of $\mathbb{R}^3$ (resp. $\mathbb{R}^2$), the characteristic function of $\mathscr{S}$ in $\mathbb{R}^3$ (resp. $\mathbb{R}^2$) will be denoted by $\chi_{\mathscr{S}}$. It is known (see e.g. [31]) that

$$\left.\begin{array}{l} \chi_{\Omega_\varepsilon^{(i)}}(x) \quad\rightharpoonup h_i(r) \quad \text{weakly in } L^2(\Omega^{(i)}), \\ \chi_{\Omega_\varepsilon^{(i)} \cap \{r=r_0\}}(x) \rightharpoonup h_i(r_0) \quad \text{weakly in } L^2(\Omega'), \end{array}\right\} \quad \text{as } \varepsilon \to 0 \quad (i = 1,\, 2). \quad (2.16)$$

With the help of identity (2.9), the first convergence in (2.16), and the fact that

$$N_\varepsilon^{(i)}(r) = 1 + \mathscr{O}(\varepsilon), \quad \varepsilon \to 0, \quad i = 1,\, 2, \quad (2.17)$$

it is easy to prove the following lemma.

Lemma 2.2 *Assume* (2.4). *Then*

$$\varepsilon \int_{\partial\Omega_\varepsilon^{(i)} \cap \{r_0 < r < r_i\}} g_\varepsilon \varphi \, d\sigma_x \longrightarrow 2 \int_{\Omega^{(i)}} g_0 \varphi \, dx \quad \text{as } \varepsilon \to 0 \;\; \forall \varphi \in H^1(\Omega^{(i)}) \; (i = 1,\, 2).$$

2.2.2 A Priori Estimates

Obviously, the bilinear forms $(\cdot, \cdot)_{1,\varepsilon}$ and $(\cdot, \cdot)_{2,\varepsilon}$ introduced in Definitions 2.1 and 2.2 define scalar products in the corresponding spaces.

Let $\| \cdot \|_{1,\varepsilon}$ be a new norm in the Sobolev space $H^1(\Omega_\varepsilon)$ produced by $(\cdot, \cdot)_{1,\varepsilon}$. Similarly to [27, Lemma 2.1], we can prove that the norms $\| \cdot \|_{1,\varepsilon}$ and $\| \cdot \|_{H^1(\Omega_\varepsilon)}$ are uniformly equivalent with respect to ε in $H^1(\Omega_\varepsilon)$, i.e., there exist positive constants ε_0, C_5, and C_6 such that

$$C_5\|\varphi\|_{1,\varepsilon} \leq \|\varphi\|_{H^1(\Omega_\varepsilon)} \leq C_6\|\varphi\|_{1,\varepsilon}, \quad \forall \varepsilon \in (0, \varepsilon_0), \ \forall \varphi \in H^1(\Omega_\varepsilon). \tag{2.18}$$

With the help of (2.12), it is easy to check (see e.g. [79, Lemma 1]) that there exist constants ε_0 and C_7 such that for all $\varepsilon \in (0, \varepsilon_0)$ and $\varphi \in H^1(\Omega_\varepsilon; \partial\Omega_\varepsilon^{(2)} \cap \{r > r_0\})$

$$\|\varphi\|_{2,\varepsilon} \leq \|\varphi\|_{H^1(\Omega_\varepsilon)} \leq C_7\|\varphi\|_{2,\varepsilon}, \tag{2.19}$$

where the norm $\| \cdot \|_{2,\varepsilon}$ is produced by $(\cdot, \cdot)_{2,\varepsilon}$.

Clearly that the functional $\mathsf{L}_{1,\varepsilon} : H^1(\Omega_\varepsilon) \mapsto \mathbb{R}$ is linear. Exploiting the Cauchy–Schwartz inequality, inequalities (2.10) and (2.13), we derive

$$|\mathsf{L}_{1,\varepsilon}(\varphi)| \leq C_8(\|f_\varepsilon\|_{L^2(\Omega_\varepsilon)} + \|q_\varepsilon\|_{L^2(\partial\Omega_0 \cap \{r<r_0\})} + \varepsilon^{\beta-1}\|g_\varepsilon\|_{H^1(\Omega^{(2)})})\|\varphi\|_{H^1(\Omega_\varepsilon)},$$

whence on the grounds of (2.18) it follows that

$$|\mathsf{L}_{1,\varepsilon}(\varphi)| \leq C_9(\|f_\varepsilon\|_{L^2(\Omega_\varepsilon)} + \|q_\varepsilon\|_{L^2(\partial\Omega_0 \cap \{r<r_0\})} + \varepsilon^{\beta-1}\|g_\varepsilon\|_{H^1(\Omega^{(2)})})\|\varphi\|_{1,\varepsilon}.$$

Using the Riesz representation theorem, we can prove that the problem (2.1) has a unique weak solution satisfying the estimate

$$\|u_\varepsilon\|_{H^1(\Omega_\varepsilon)} \leq C_{10}(\|f_\varepsilon\|_{L^2(\Omega_\varepsilon)} + \|q_\varepsilon\|_{L^2(\partial\Omega_0 \cap \{r<r_0\})} + \varepsilon^{\beta-1}\|g_\varepsilon\|_{H^1(\Omega^{(2)})}). \tag{2.20}$$

Similarly with the help of (2.19), we show that there exists a unique weak solution to the problem (2.2), which satisfies the estimate

$$\|v_\varepsilon\|_{H^1(\Omega_\varepsilon)} \leq C_{11}(\|f_\varepsilon\|_{L^2(\Omega_\varepsilon)} + \|q_\varepsilon\|_{L^2(\partial\Omega_0 \cap \{r<r_0\})} + \|g_\varepsilon\|_{H^1(\Omega^{(1)})}). \tag{2.21}$$

2.3 Convergence Theorems

2.3.1 Convergence Theorem for Problem (2.1)

For every ε and for every function $\varphi \in L^2(\Omega_\varepsilon^{(i)})$, we set

$$\widetilde{\varphi}^{(i)}(x) = \begin{cases} \varphi(x), & x \in \Omega_\varepsilon^{(i)}, \\ 0, & x \in \Omega^{(i)} \backslash \Omega_\varepsilon^{(i)} \end{cases} \quad (i = 1, 2).$$

Obviously, if $\varphi \in H^1(\Omega_\varepsilon; \partial\Omega_\varepsilon^{(2)} \cap \{r > r_0\})$, then $\widetilde{\varphi}^{(2)} \in H^1(\Omega^{(2)})$.

Let us introduce a space of Sobolev space $\widetilde{\mathbf{H}}$ of anisotropic multi-sheeted functions. A multi-sheeted function

$$\mathbf{p}(x) := (p_0, \ p_1, \ p_2) = \begin{cases} p_0(x), & x \in \Omega_0, \\ p_1(x), & x \in \Omega^{(1)}, \\ p_2(x), & x \in \Omega^{(2)}, \end{cases}$$

belongs to $\widetilde{\mathbf{H}}$ if $p_0 \in H^1(\Omega_0)$, $p_i \in \widetilde{H}^1(\Omega^{(i)})$, $i = 1, 2$, and $p_0|_{\Omega'} = p_1|_{\Omega'} = p_2|_{\Omega'}$.

Obviously, the space $\widetilde{\mathbf{H}}$ continuously and densely embedded in the Hilbert space $\mathbf{L}$ of multi-sheeted functions whose components belong to the corresponding L^2-spaces, i.e., $\mathbf{p} \in \mathbf{L}$ if $p_0 \in L^2(\Omega_0)$, $p_i \in L^2(\Omega^{(i)})$, $i = 1, 2$. The scalar product in $\mathbf{L}$ defined as follows:

$$(\mathbf{p}, \ \mathbf{q})_{\mathbf{L}} = \int_{\Omega_0} p_0 q_0 \, dx + \sum_{i=1}^{2} \int_{\Omega^{(i)}} h_i(r) p_i q_i \, dx,$$

and in $\widetilde{\mathbf{H}}$ by the formula

$$(\mathbf{p}, \ \mathbf{q})_{\widetilde{\mathbf{H}}} = \int_{\Omega_0} \nabla_x p_0 \cdot \nabla_x q_0 \, dx + \sum_{i=1}^{2} \int_{\Omega^{(i)}} h_i(r) \nabla_{\tilde{x}} p_i \cdot \nabla_{\tilde{x}} q_i \, dx$$

$$+ 2\delta_{\alpha,1} k_1 \int_{\Omega^{(1)}} p_1 q_1 \, dx + 2k_2 \int_{\Omega^{(2)}} p_2 q_2 \, dx + k_2 h_2(r_2) \int_{\partial\Omega^{(2)} \cap \{r=r_2\}} p_2 q_2 \, d\sigma_x,$$

where $\mathbf{p} = (p_0, \ p_1, \ p_2)$, $\mathbf{q} = (q_0, \ q_1, \ q_2)$, $\delta_{i,j}$ is the Kronecker symbol.

Theorem 2.1 *Assume* (2.3)–(2.5). *Then for the weak solution u_ε to problem* (2.1) *the following limit relations hold:*

$$\left. \begin{array}{l} u_\varepsilon \rightharpoonup u_0 \quad \text{weakly in } H^1(\Omega_0), \\ \widetilde{u}_\varepsilon^{(1)} \rightharpoonup h_1 u_1 \text{ weakly in } L^2(\Omega^{(1)}), \\ \widetilde{u}_\varepsilon^{(2)} \rightharpoonup h_2 u_2 \text{ weakly in } L^2(\Omega^{(2)}) \end{array} \right\} \quad \text{as } \varepsilon \to 0, \qquad (2.22)$$

where the function $\mathbf{u} = (u_0, \ u_1, \ u_2) \in \widetilde{\mathbf{H}}$ is a weak solution to the problem

$$\begin{cases} -\Delta_x u_0(x) = f_0(x), & x \in \Omega_0, \\ \partial_\nu u_0(x) = q_0(x), & x \in \partial\Omega_0 \cap \{r < r_0\}, \\ -\operatorname{div}_{\tilde{x}}(h_1(r)\nabla_{\tilde{x}} u_1(x)) + 2k_1\delta_{\alpha,1} u_1(x) \\ \qquad = h_1(r)f_0(x) + 2\delta_{\beta,1} g_0(x), & x \in \Omega^{(1)}, \\ \partial_\nu u_1(x) = 0, & x \in \partial\Omega^{(1)} \cap \{r = r_1\}, \\ -\operatorname{div}_{\tilde{x}}(h_2(r)\nabla_{\tilde{x}} u_2(x)) + 2k_2 u_2(x) \\ \qquad = h_2(r)f_0(x) + 2\delta_{\beta,1} g_0(x), & x \in \Omega^{(2)}, \\ \partial_\nu u_2(x) + k_2 u_2(x) = 0, & x \in \partial\Omega^{(2)} \cap \{r = r_2\}, \\ u_0(x) = u_1(x) = u_2(x), & x \in \Omega', \\ \partial_r u_0(x) = \sum\limits_{i=1}^{2} h_i(r_0)\partial_r u_i(x), & x \in \Omega'. \end{cases} \tag{2.23}$$

Besides, the following energy convergence takes place:

$$E_{1,\varepsilon}(u_\varepsilon) := (u_\varepsilon, u_\varepsilon)_{1,\varepsilon} \longrightarrow (\mathbf{u}, \mathbf{u})_{\widetilde{\mathbf{H}}} =: E_1(\mathbf{u}) \quad \varepsilon \to 0. \tag{2.24}$$

Problem (2.23) is called a *homogenized problem* for problem (2.1).

Definition 2.3 A multi-sheeted function $\mathbf{u} \in \widetilde{\mathbf{H}}$ is called a weak solution to problem (2.23) if

$$(\mathbf{u}, \mathbf{p})_{\widetilde{\mathbf{H}}} = \mathsf{L}_1(\mathbf{p}) \quad \forall \mathbf{p} \in \widetilde{\mathbf{H}}, \tag{2.25}$$

where

$$\mathsf{L}_1(\mathbf{p}) = \int_{\Omega_0} f_0 p_0 \, dx + \int_{\partial\Omega_0 \cap \{r < r_0\}} q_0 p_0 \, d\tilde{x} + \sum_{i=1}^{2} \int_{\Omega^{(i)}} (h_i(r) f_0 + 2\delta_{\beta,1} g_0) p_i \, dx.$$

It is easy to verify that L_1 is a continuous functional in $\widetilde{\mathbf{H}}$. Then the Riesz theorem supplies the existence and uniqueness of a weak solution to problem (2.23).

Proof 1. Relations (2.20), (2.13), (2.3), (2.4), and (2.5) imply the uniform (with respect to ε) boundedness of the following quantities: $\|u_\varepsilon\|_{H^1(\Omega_0)}$, $\|\widetilde{u}_\varepsilon^{(i)}\|_{L^2(\Omega^{(i)})}$, $\|\widetilde{\partial_{x_k} u_\varepsilon}^{(i)}\|_{L^2(\Omega^{(i)})}$, $\|\widetilde{u}_\varepsilon|_{\partial\Omega_\varepsilon^{(2)} \cap \{r=r_2\}}\|_{L^2(\partial\Omega^{(2)} \cap \{r=r_2\})})$ $(i = 1, 2; \ k = 1, 2, 3)$. Here

$$\widetilde{\varphi}(x)|_{\partial\Omega_\varepsilon^{(i)} \cap \{r=r_i\}} := \begin{cases} \varphi(x)|_{\partial\Omega_\varepsilon^{(i)} \cap \{r=r_i\}}, & x \in \partial\Omega_\varepsilon^{(i)} \cap \{r = r_i\}, \\ 0, & x \in (\partial\Omega^{(i)} \cap \{r = r_i\}) \setminus (\partial\Omega_\varepsilon^{(i)} \cap \{r = r_i\}) \end{cases}$$

$\varphi \in H^1(\Omega_\varepsilon^{(i)})$, $i = 1, 2$., Therefore, there exists a subsequence $\{\varepsilon'\} \subset \{\varepsilon\}$ (again denoted by ε) such that

$$\begin{aligned}
u_\varepsilon &\rightharpoonup u_0 & \text{weakly in } H^1(\Omega_0),\\
\widetilde{u}_\varepsilon^{(i)} &\rightharpoonup \widetilde{u}_i := h_i u_i & \text{weakly in } L^2(\Omega^{(i)}),\\
\widetilde{\partial_{x_k} u}_\varepsilon^{(i)} &\rightharpoonup \widetilde{u}_{i,k} := h_i u_{i,k} & \text{weakly in } L^2(\Omega^{(i)}),\\
\widetilde{u}_\varepsilon|_{\partial\Omega_\varepsilon^{(2)}\cap\{r=r_2\}} &\rightharpoonup \widetilde{u}_{r_2} := h_2(r_2)u_{r_2} & \text{weakly in } L^2(\partial\Omega^{(2)}\cap\{r=r_2\})
\end{aligned} \tag{2.26}$$

as $\varepsilon \to 0$, the functions u_0, u_i, $u_{i,k}$, u_{r_2} will be defined later; $i = 1, 2;\ k = 1, 2, 3$.

2. At first we find $u_{i,2}$. Consider the following test functions:

$$\Phi_1(x) = \begin{cases} 0, & x \in \Omega_0 \cup \Omega_\varepsilon^{(2)}, \\ \varepsilon Y_1(\frac{x_2}{\varepsilon})\varphi_1(x), & x \in \Omega_\varepsilon^{(1)}, \end{cases}$$

$$\Phi_2(x) = \begin{cases} 0, & x \in \Omega_0 \cup \Omega_\varepsilon^{(1)}, \\ \varepsilon Y_2(\frac{x_2}{\varepsilon})\varphi_2(x), & x \in \Omega_\varepsilon^{(2)}, \end{cases}$$

where $\varphi_i \in C_0^\infty(\Omega^{(i)})$, $i = 1,\ 2$, are arbitrary functions and the function Y_2 is defined in (2.9). Obviously, $\Phi_i \in H^1(\Omega_\varepsilon)$ and

$$\nabla_x \Phi_i(x) = (0,\ -\varphi_i(x),\ 0) + \varepsilon Y_i\left(\frac{x_2}{\varepsilon}\right)\nabla_x\varphi_i(x), \quad x \in \Omega_\varepsilon^{(i)},\ i = 1,\ 2.$$

Substituting the function Φ_1 into identity (2.6), we get

$$\left|\int_{\Omega_\varepsilon^{(1)}} \partial_{x_2} u_\varepsilon \varphi_1\, dx\right| \le \varepsilon \left(\int_{\Omega_\varepsilon^{(1)}} |\nabla_x u_\varepsilon \cdot \nabla_x \varphi_1|\, dx + \varepsilon^\alpha k_1 \int_{\partial\Omega_\varepsilon^{(1)}\cap\{r>r_0\}} |u_\varepsilon\varphi_1|\, d\sigma_x \right.$$
$$\left. + \int_{\Omega_\varepsilon^{(1)}} |f_\varepsilon\varphi_1|\, dx + \varepsilon^\beta \int_{\partial\Omega_\varepsilon^{(1)}\cap\{r>r_0\}} |g_\varepsilon\varphi_1|\, d\sigma_x\right).$$

Passing to the limit in this inequality and taking into account (2.26), (2.20), (2.10), (2.13), (2.3), (2.4), and that $\alpha,\ \beta \ge 1$, we obtain

$$\int_{\Omega^{(1)}} h_1(r)u_{1,2}\varphi_1\, dx = 0 \quad \forall\, \varphi_1 \in C_0^\infty(\Omega^{(1)}),$$

whence $u_{1,2} = 0$ a.e. in $\Omega^{(1)}$. Similarly, using Φ_2, we get that $u_{2,2} = 0$ a.e. in $\Omega^{(2)}$.

Now integrating by parts and taking into account (2.8), we derive the identity

$$\int_{\Omega_\varepsilon^{(1)}} \partial_{x_k} u_\varepsilon \psi\, dx = -\int_{\Omega_\varepsilon^{(1)}} u_\varepsilon \partial_{x_k}\psi\, dx - \frac{\varepsilon}{2}\int_{\partial\Omega_\varepsilon^{(1)}\cap\{r_0<r<r_1\}} \left(N_\varepsilon^{(1)}(r)\right)^{-1}\partial_{x_k} h_1 u_\varepsilon \psi\, d\sigma_x,$$

where $k = 1,\ 3$ and $\psi \in C_0^\infty(\Omega^{(1)})$. Using (2.9) and the zero extension operators, we rewrite this identity as follows:

$$\int_{\Omega^{(1)}} \widetilde{\partial_{x_k} u}_\varepsilon^{(1)} \psi \, dx = - \int_{\Omega^{(1)}} \widetilde{u}_\varepsilon^{(1)} (\partial_{x_k} \psi + \partial_{x_k} \ln h_1(r) \psi) \, dx$$

$$+ \varepsilon \int_{\Omega_\varepsilon^{(1)}} Y_1 \left(\frac{x_2}{\varepsilon} \right) \partial_{x_k} \ln h_1(r) \partial_{x_2} (u_\varepsilon \psi) \, dx, \quad k = 1, 3. \quad (2.27)$$

Evidently the last integral in (2.27) vanishes as $\varepsilon \to 0$. The limits of the rest ones can be found with the help of (2.26). As a result, we have

$$\int_{\Omega^{(1)}} u_{1,k} h_1(r) \psi \, dx = - \int_{\Omega^{(1)}} u_1 \partial_{x_k} (h_1(r) \psi) \, dx \quad \forall \psi \in C_0^\infty(\Omega^{(1)}),$$

whence it follows the existence of the weak derivatives $\partial_{x_k} u_1$ and $u_{1,k} = \partial_{x_k} u_1$ a.e. $\Omega^{(1)}$, $k = 1, 3$. Similarly we can show that $\partial_{x_k} u_2 = u_{2,k}$ a.e. in $\Omega^{(2)}$, $k = 1, 3$.

Let us find u_{r_2}. Direct calculations give that

$$\partial_{x_1} (r^{-1} \cos \theta) + \partial_{x_3} (r^{-1} \sin \theta) = 0. \quad (2.28)$$

Consider an arbitrary function $\psi \in C^\infty(\overline{\Omega^{(2)}})$ such that $\psi|_{\Omega'} = 0$. Integrating by parts in $\Omega_\varepsilon^{(2)}$ and bearing in mind (2.8) and (2.28), we derive the identity

$$\int_{\Omega_\varepsilon^{(2)}} r^{-1} \partial_r u_\varepsilon \psi \, dx = \int_{\Omega_\varepsilon^{(2)}} r^{-1} (\partial_{x_1} u_\varepsilon \cos \theta + \partial_{x_3} u_\varepsilon \sin \theta) \psi \, dx$$

$$= - \int_{\Omega_\varepsilon^{(2)}} r^{-1} u_\varepsilon \partial_r \psi \, dx + r_2^{-1} \int_{\partial \Omega_\varepsilon^{(2)} \cap \{r=r_2\}} u_\varepsilon \psi \, d\sigma_x$$

$$- \frac{\varepsilon}{2} \int_{\partial \Omega_\varepsilon^{(2)} \cap \{r_0 < r < r_2\}} \left(N_\varepsilon^{(2)}(r) \right)^{-1} r^{-1} h_2'(r) u_\varepsilon \psi \, d\sigma_x.$$

With the help of (2.9) and the zero extensions, we rewrite it as

$$r_2^{-1} \int_{\partial \Omega^{(2)} \cap \{r=r_2\}} \widetilde{u}_\varepsilon |_{\partial \Omega_\varepsilon^{(2)} \cap \{r=r_2\}} \psi \, d\sigma_x$$

$$= \int_{\Omega^{(2)}} r^{-1} (\widetilde{\partial_r u}_\varepsilon^{(1)} \psi + \widetilde{u}_\varepsilon^{(1)} \partial_r \psi + (\ln h_2(r))' \widetilde{u}_\varepsilon^{(1)} \psi) \, dx$$

$$- \varepsilon \int_{\Omega_\varepsilon^{(2)}} Y_2 \left(\frac{x_2}{\varepsilon} \right) r^{-1} (\ln h_2(r))' \partial_{x_2} (u_\varepsilon \psi) \, dx. \quad (2.29)$$

Evidently, the last integral vanishes in (2.29) as $\varepsilon \to 0$. The limits of the other ones can be found with (2.26). As a result, we get

$$r_2^{-1} h_2(r_2) \int_{\partial \Omega^{(2)} \cap \{r=r_2\}} u_{r_2} \psi \, d\sigma_x = \int_{\Omega^{(2)}} r^{-1} \partial_r (h_2(r) u_2 \psi) \, dx.$$

After integrating by parts, we obtain

$$\int_{\partial\Omega^{(2)}\cap\{r=r_2\}} u_{r_2}\psi\,d\sigma_x = \int_{\partial\Omega^{(2)}\cap\{r=r_2\}} u_2\psi\,d\sigma_x \quad \forall\,\psi \in C^\infty(\overline{\Omega^{(2)}}),\ \psi|_{\Omega'}=0,$$

whence it follows that $u_{r_2} = u_2|_{\partial\Omega^{(2)}\cap\{r=r_2\}}$ a..e. in $\partial\Omega^{(2)}\cap\{r=r_2\}$.

3. Let us find conjugation conditions on the joint zone. The first relation in (2.26) and compactness of the trace operator imply that

$$u_\varepsilon|_{\Omega'} \longrightarrow u_0|_{\Omega'} \quad \text{strongly in } L^2(\Omega'). \tag{2.30}$$

Similarly as we have deduced (2.29), we can prove that the following identity:

$$-r_0^{-1}\int_{\Omega'} \chi_{\Omega_\varepsilon^{(1)}\cap\{r=r_0\}} u_\varepsilon\psi\,d\sigma_x = \int_{\Omega^{(1)}} r^{-1}(\widetilde{\partial_r u}_\varepsilon^{(1)}\psi + \widetilde{u}_\varepsilon^{(1)}\partial_r\psi + (\ln h_1(r))'\widetilde{u}_\varepsilon^{(1)}\psi)\,dx$$
$$- \varepsilon\int_{\Omega_\varepsilon^{(1)}} Y_1\left(\frac{x_2}{\varepsilon}\right) r^{-1}(\ln h_1(r))'\partial_{x_2}(u_\varepsilon\psi)\,dx \tag{2.31}$$

for any $\psi \in C^\infty(\overline{\Omega^{(1)}})$, $\psi|_{\partial\Omega^{(1)}\cap\{r=r_1\}}=0$. Utilizing the second relation in (2.16), (2.30) and (2.26), we can pass to the limit in (2.31). As a result, we get

$$-\frac{h_1(r_0)}{r_0}\int_{\Omega'} u_0\psi\,d\sigma_x = \int_{\Omega^{(1)}} r^{-1}\partial_r(h_1(r)u_1\psi)\,dx = -\frac{h_1(r_0)}{r_0}\int_{\Omega'} u_1\psi\,d\sigma_x,$$

whence $u_0|_{\Omega'}=u_1|_{\Omega'}$ a.e. in Ω'.

Repeating the same assertions for the thin discs from the second level, we obtain

$$u_0|_{\Omega'}=u_1|_{\Omega'}=u_2|_{\Omega'} \quad \text{a.e. in } \Omega'. \tag{2.32}$$

On the grounds of (2.32) and results obtained in the second item, we can claim that the multi-sheeted function $\mathbf{u}=(u_0,u_1,u_2)$ belongs to $\widetilde{\mathbf{H}}$.

4. Consider a multi-sheeted function $\mathbf{p}=(p_0,p_1,p_2)$, where $p_0 \in C^\infty(\overline{\Omega}_0)$, $p_i \in C^\infty(\overline{\Omega^{(i)}})$, $i=1,2$, are arbitrary functions such that $p_0|_{\Omega'}=p_1|_{\Omega'}=p_2|_{\Omega'}$. Then the function

$$\Phi(x) = \mathbf{p}|_{\overline{\Omega}_\varepsilon}(x) := \begin{cases} p_0(x), & x \in \Omega_0, \\ p_1(x), & x \in \Omega_\varepsilon^{(1)}, \\ p_2(x), & x \in \Omega_\varepsilon^{(2)}. \end{cases}$$

belongs to $H^1(\Omega_\varepsilon)$. Substituting Φ in (2.6) and utilizing the zero extensions, we get

$$\int_{\Omega_0} \nabla_x u_\varepsilon \cdot \nabla_x p_0\,dx + \sum_{i=1}^{2}\int_{\Omega^{(i)}} \widetilde{\nabla_x u}_\varepsilon^{(i)} \cdot \nabla_x p_i\,dx + \varepsilon^\alpha k_1\int_{\partial\Omega_\varepsilon^{(1)}\cap\{r>r_0\}} u_\varepsilon p_1\,d\sigma_x$$

$$\varepsilon k_2 \int_{\partial\Omega_\varepsilon^{(2)}\cap\{r_0<r<r_2\}} u_\varepsilon p_2 \, d\sigma_x + k_2 \int_{\partial\Omega^{(2)}\cap\{r=r_2\}} \widetilde{u}_\varepsilon|_{\partial\Omega_\varepsilon^{(2)}\cap\{r=r_2\}} p_2 \, d\sigma_x$$

$$= \int_{\Omega_0} f_\varepsilon p_0 \, dx + \sum_{i=1}^{2} \left(\int_{\Omega^{(i)}} \chi_{\Omega^{(i)}} f_\varepsilon p_i \, dx + \varepsilon^\beta \int_{\partial\Omega_\varepsilon^{(i)}\cap\{r_0<r<r_i\}} g_\varepsilon p_i \, d\sigma_x \right)$$

$$+\varepsilon^\beta \int_{\partial\Omega_\varepsilon^{(1)}\cap\{r=r_1\}} g_\varepsilon p_1 \, d\sigma_x + \int_{\partial\Omega_0\cap\{r<r_0\}} q_\varepsilon p_0 \, d\widetilde{x},$$

where $\widetilde{\nabla_x u}_\varepsilon^{(i)} = (\widetilde{\partial_{x_1} u}_\varepsilon^{(i)}, \widetilde{\partial_{x_2} u}_\varepsilon^{(i)}, \widetilde{\partial_{x_3} u}_\varepsilon^{(i)})$, $i = 1, 2$. With the help of (2.26), (2.13), (2.3), (2.4), (2.5), (2.16), and Lemma 2.2, this equality is transformed ($\varepsilon \to 0$) in (2.25) for the function $\mathbf{u}$. By approximation the obtained identity holds for any function $\mathbf{p} \in \widetilde{\mathbf{H}}$. Hence, the function $\mathbf{u}$ is the unique weak solution to problem (2.23).

5. Due to the uniqueness of this solution, the above assertions are true for any subsequence $\{\varepsilon'\}$ chosen at the beginning of the proof. Thus the limits (2.22) hold.

6. It remains to prove the convergence (2.24) of the energy integral $E_{1,\varepsilon}(u_\varepsilon)$ to the energy integral $E_1(\mathbf{u})$ of the homogenized problem. Using (2.6), we get

$$E_{1,\varepsilon}(u_\varepsilon) = \mathsf{L}_{1,\varepsilon}(u_\varepsilon) = \int_{\Omega_0} f_\varepsilon u_\varepsilon \, dx + \varepsilon^\beta \int_{\partial\Omega_\varepsilon^{(1)}\cap\{r=r_1\}} g_\varepsilon u_\varepsilon \, d\sigma_x$$

$$+ \int_{\partial\Omega_0\cap\{r<r_0\}} q_\varepsilon u_\varepsilon \, d\widetilde{x} + \sum_{i=1}^{2} \left(\int_{\Omega^{(i)}} f_\varepsilon \widetilde{u}_\varepsilon^{(i)} \, dx + \varepsilon^\beta \int_{\partial\Omega_\varepsilon^{(i)}\cap\{r_0<r<r_i\}} g_\varepsilon u_\varepsilon \, d\sigma_x \right).$$
$$\tag{2.33}$$

Taking into account (2.9), (2.26), (2.4), and compact embedding $H^1(\Omega^{(i)}) \subset L^2(\Omega^{(i)})$, similarly as in Lemma 2.2 we prove

$$\varepsilon \int_{\partial\Omega_\varepsilon^{(i)}\cap\{r_0<r<r_i\}} g_\varepsilon u_\varepsilon \, d\sigma_x \longrightarrow 2 \int_{\Omega^{(i)}} g_0 u_i \, dx \quad \text{as } \varepsilon \to 0, \; i = 1, 2.$$

The limits of the rest of the summands in (2.33) can be found with the help of (2.3), (2.4), (2.5), (2.26), and (2.13). Thus,

$$\lim_{\varepsilon\to 0} E_{1,\varepsilon}(u_\varepsilon) = \mathsf{L}_1(\mathbf{u}) = E_1(\mathbf{u}).$$

2.3.2   Convergence Theorem for Problem (2.2)

Let us introduce Sobolev spaces $H^1(\Omega_0; \Omega') = \{\varphi \in H^1(\Omega_0) : \; \varphi|_{\Omega'} = 0\}$ and $\widetilde{H}^1(\Omega^{(i)}; \Omega') = \{\varphi \in \widetilde{H}^1(\Omega^{(i)}) : \; \varphi|_{\Omega'} = 0\}$, $i = 1, 2$, with the scalar products

$$(u, v)_{2,0} = \int_{\Omega_0} \nabla_x u \cdot \nabla_x v \, dx \quad \forall u, v \in H^1(\Omega_0; \Omega')$$

$$(u, v)_{2,i} = \int_{\Omega^{(i)}} h_i(r) \nabla_{\tilde{x}} u \cdot \nabla_{\tilde{x}} v \, dx \quad \forall u, v \in \tilde{H}^1(\Omega^{(i)}; \Omega').$$

Theorem 2.2 *Assume (2.3)–(2.5). Then for the weak solution v_ε to problem (2.2) the following limit relations hold:*

$$\left. \begin{array}{ll} v_\varepsilon \rightharpoonup v_0 & \text{weakly in } H^1(\Omega_0), \\ \tilde{v}_\varepsilon^{(1)} \rightharpoonup h_1 v_1 & \text{weakly in } L^2(\Omega^{(1)}), \\ \tilde{v}_\varepsilon^{(2)} \rightharpoonup 0 & \text{weakly in } H^1(\Omega^{(2)}) \end{array} \right\} \quad \text{as } \varepsilon \to 0, \tag{2.34}$$

where $v_0 \in H^1(\Omega_0; \Omega')$ is a weak solution to the problem

$$\begin{cases} -\Delta_x v_0(x) = f_0(x), & x \in \Omega_0, \\ \partial_\nu v_0(x) = q_0(x), & x \in \partial\Omega_0 \cap \{r < r_0\}, \\ v_0(x) = 0, & x \in \Omega', \end{cases} \tag{2.35}$$

and $v_1 \in \tilde{H}^1(\Omega^{(1)}; \Omega')$ is a weak solution to the problem

$$\begin{cases} -\mathrm{div}_{\tilde{x}}\big(h_1(r)\nabla_{\tilde{x}} v_1(x)\big) = h_1(r) f_0(x) + 2g_0(x), & x \in \Omega^{(1)}, \\ \partial_\nu v_1(x) = 0, & x \in \partial\Omega^{(1)} \cap \{r = r_1\}, \quad (2.36) \\ v_1(x) = 0, & x \in \Omega'. \end{cases}$$

Besides, the following energy convergence (as $\varepsilon \to 0$) takes place:

$$E_{2,\varepsilon}(v_\varepsilon) := (v_\varepsilon, v_\varepsilon)_{2,\varepsilon} \longrightarrow (v_0, v_0)_{2,0} + (v_1, v_1)_{2,1} =: E_{2,0}(v_0) + E_{2,1}(v_1). \tag{2.37}$$

Both problems (2.35) and (2.36) form the *homogenized problem* for problem (2.2).

Definition 2.4 A function $v \in H^1(\Omega_0; \Omega')$ is called a weak solution to problem (2.35) if the integral identity

$$(v, \varphi)_{2,0} = \mathsf{L}_{2,0}(\varphi) := \int_{\Omega_0} f_0 \varphi \, dx + \int_{\partial\Omega_0 \cap \{r < r_0\}} q_0 \varphi \, d\tilde{x} \tag{2.38}$$

holds for all $\varphi \in H^1(\Omega_0; \Omega')$.

Definition 2.5 A function $v \in \tilde{H}^1(\Omega^{(1)}; \Omega')$ is called a weak solution to problem (2.36) if the identity

$$(v, \varphi)_{2,1} = \mathsf{L}_{2,1}(\varphi) := \int_{\Omega^{(1)}} (h_1(r) f_0 + 2g_0)\varphi \, dx \tag{2.39}$$

is valid for every $\varphi \in \widetilde{H}^1(\Omega^{(1)}; \Omega')$.

Using the Riesz representation theorem, it is easy to check that these problems have unique weak solutions.

Proof **1.** Estimate (2.21) and (2.3)–(2.5) imply that the quantities

$$\|v_\varepsilon\|_{H^1(\Omega_0)}, \quad \|\widetilde{v}_\varepsilon^{(2)}\|_{H^1(\Omega^{(2)})}, \quad \|\widetilde{v}_\varepsilon^{(1)}\|_{L^2(\Omega^{(1)})}, \quad \|\widetilde{\partial_{x_k} v}_\varepsilon^{(1)}\|_{L^2(\Omega^{(1)})}, \ k = 1, 2, 3,$$

are uniformly bounded with respect to ε. Hence there exists a subsequence $\{\varepsilon'\} \subset \{\varepsilon\}$ (again denoted by $\{\varepsilon\}$) such that

$$\left.\begin{array}{ll} v_\varepsilon \rightharpoonup v_0 & \text{weakly in } H^1(\Omega_0), \\ \widetilde{v}_\varepsilon^{(1)} \rightharpoonup \widetilde{v}_1 := h_1 v_1 & \text{weakly in } L^2(\Omega^{(1)}), \\ \widetilde{\partial_{x_k} v}_\varepsilon^{(1)} \rightharpoonup \widetilde{v}_{1,k} := h_1 v_{1,k} & \text{weakly in } L^2(\Omega^{(1)}) \\ \widetilde{v}_\varepsilon^{(2)} \rightharpoonup v_2 & \text{weakly in } H^1(\Omega^{(2)}), \end{array}\right\} \quad \text{as } \varepsilon \to 0 \ (k = 1, 2, 3),$$

(2.40)

where the functions v_0, v_1, $v_{1,k}$, v_2 will be defined later.

2. From (2.12) and (2.21) it follows that

$$\|\widetilde{v}_\varepsilon^{(2)}\|_{L^2(\Omega^{(2)})}^2 \le c_0\varepsilon^2 \int_{\Omega_\varepsilon^{(2)}} |\nabla_x v_\varepsilon|^2 \, dx \le c_0\varepsilon^2,$$

whence we deduce that $v_2 = 0$ a.e. in $\Omega^{(2)}$.

Similarly as in steps 2 and 3 in the proof of Theorem 2.1 we can show that

- $v_{1,2} = 0$ a.e. in $\Omega^{(1)}$;
- there exist weak derivatives $\partial_{x_k} v_1$, $k = 1, 3$, and $\partial_{x_k} v_1 = v_{1,k}$ a.e. in $\Omega^{(1)}$;
- $v_0|_{\Omega'} = v_1|_{\Omega'} = v_2|_{\Omega'}$ a.e. in Ω'.

Since $v_2 = 0$ a.e. in $\Omega^{(2)}$, $v_0|_{\Omega'} = v_1|_{\Omega'} = 0$ a.e. in Ω'.

3. Let us consider arbitrary functions $\varphi_0 \in C^\infty(\overline{\Omega_0})$ and $\varphi_1 \in C^\infty(\overline{\Omega^{(1)}})$ such that $\varphi_0 = \varphi_1 = 0$ in some neighborhood of Ω'. With the help of these functions, we define a function

$$\Phi(x) = \begin{cases} \varphi_0(x), & x \in \Omega_0, \\ \varphi_1(x), & x \in \Omega_\varepsilon^{(1)}, \\ 0, & x \in \Omega_\varepsilon^{(2)}. \end{cases}$$

Obviously, $\Phi \in H^1(\Omega_\varepsilon; \partial\Omega_\varepsilon^{(2)} \cap \{r > r_0\})$. Integral identity (2.7) with the test function Φ looks like

$$\int_{\Omega_0} \nabla_x v_\varepsilon \cdot \nabla_x \varphi_0 \, dx + \int_{\Omega^{(1)}} \widetilde{\nabla_x v}_\varepsilon^{(1)} \cdot \nabla_x \varphi_1 \, dx = \int_{\Omega_0} f_\varepsilon \varphi_0 \, dx + \int_{\Omega^{(1)}} \chi_{\Omega^{(1)}} f_\varepsilon \varphi_1 \, dx$$

$$+ \varepsilon \int_{\partial\Omega_\varepsilon^{(1)} \cap \{r > r_0\}} g_\varepsilon \varphi_1 \, d\sigma_x + \int_{\partial\Omega_0 \cap \{r < r_0\}} q_\varepsilon \varphi_0 \, d\tilde{x}.$$

Passing to the limit in the last identity and taking into account (2.40), (2.3), (2.4), (2.5), (2.16), (2.13), and Lemma 2.2, we get

$$(v_0, \varphi_0)_{2,0} + (v_1, \varphi_1)_{2,1} = \mathsf{L}_{2,0}(\varphi_0) + \mathsf{L}_{2,1}(\varphi_1).$$

Evidently, this identity is equivalent to the following two identities: (2.38) (for v_0 and for all $\varphi_0 \in C^\infty(\overline{\Omega_0})$, $\varphi_0 = 0$ in some neighborhood of Ω') and (2.39) (for v_1 and for all $\varphi_1 \in C^\infty(\overline{\Omega^{(1)}})$, $\varphi_1 = 0$ in some neighborhood of Ω'). By the closure, identity (2.38) holds for every $\varphi_0 \in H^1(\Omega_0; \Omega')$, and identity (2.39) holds for every $\varphi_1 \in \widetilde{H}^1(\Omega^{(1)}; \Omega')$. Thus, v_0 is the unique weak solution to problem (2.35) and v_1 is the unique weak solution to problem (2.36).

Due to the uniqueness of these solutions, the above assertions are true for any subsequence $\{\varepsilon'\}$ chosen at the beginning of the proof. Thus the limits (2.34) hold.

4. In order to prove (2.37), we first note from (2.7) that

$$E_{2,\varepsilon}(v_\varepsilon) = \int_{\Omega_\varepsilon} |\nabla_x v_\varepsilon|^2 \, dx$$

$$= \int_{\Omega_0} f_\varepsilon v_\varepsilon \, dx + \sum_{i=1}^{2} \int_{\Omega^{(i)}} f_\varepsilon \widetilde{v}_\varepsilon^{(i)} \, dx + \varepsilon \int_{\partial\Omega_\varepsilon^{(1)} \cap \{r > r_0\}} g_\varepsilon v_\varepsilon \, d\sigma_x + \int_{\partial\Omega_0 \cap \{r < r_0\}} q_\varepsilon v_\varepsilon \, d\tilde{x}.$$

Passing to the limit as $\varepsilon \to 0$ in this equality and taking into account (2.3), (2.4), (2.5), (2.40), (2.16), (2.13), (2.9), we deduce

$$\lim_{\varepsilon \to 0} E_{2,\varepsilon}(v_\varepsilon) = \mathsf{L}_{2,0}(v_0) + \mathsf{L}_{2,1}(v_1).$$

Taking (2.38) and (2.39) into account, we get (2.37). The theorem is proved.

2.4   Conclusions to this Chapter

The results obtained in this chapter show that boundary conditions on the surfaces of the thin discs, geometric structure of the discs, the parameters α and β significantly influence the asymptotic behavior (as $\varepsilon \to 0$) of solutions to BVPs in thick multilevel junctions of type 3:2:2.

1. In case of the alternating Robin boundary conditions (problem (2.1)), the solution to the corresponding homogenized problem (2.23) is the multi-sheeted function $\mathbf{u} = (u_0, u_1, u_2)$ whose components are the first terms of the asymptotics for the solution u_ε to problem (2.1). The homogenized problem (2.23) contains three different differential equations, namely the Poisson equation in the junction's body Ω_0, and two homogenized differential equations with respect to only two variables x_1 and x_3 in the domains $\Omega^{(1)}$ and $\Omega^{(2)}$, respectively (the variable x_2 is involved here as a parameter). This is a consequence of the type of the thick junction Ω_ε. In addition,

- the functions h_1 and h_2 appear as coefficients in the differential equations in $\Omega^{(1)}$ and $\Omega^{(2)}$ as well as in the second conjugation condition on Ω':

$$\partial_r u_0 = h_1(r_0)\partial_r u_1 + h_2(r_0)\partial_r u_2$$

(recall that these functions describe relative thickness of the thin discs);
- the Robin boundary conditions are transformed into the zero-order terms $2\delta_{\alpha,1}k_1 u_1$, $2k_2 u_2$, and $2\delta_{\beta,1}g_0$ of the homogenized differential equations in $\Omega^{(1)}$ and $\Omega^{(2)}$. These terms show the influence of the parameters α and β. If $\alpha > 1$, then the summand $2\delta_{\alpha,1}k_1 u_1$ vanishes. From physical point of view, this means that thermal conductivity on the surfaces of the thin discs from the first level is so small that we can neglect the heat exchange on this part of the boundary. If $\beta > 1$, then the temperature outside can be neglected.

2. In case of the alternating Neumann and Dirichlet boundary conditions (problem (2.2)) the zero extension $\widetilde{v}_\varepsilon^{(2)}$ of the solution tends to zero (see the last limit in (2.34)). As a result, the starting problem (2.2) is split (as $\varepsilon \to 0$) into two independent problems (2.35) and (2.36) that form together the homogenized problem for problem (2.2). This union is justified by the energy convergence (2.37) as well. In addition, next terms of the asymptotics for v_ε depend on the solutions to problems (2.35) and (2.36).
3. The convergence of the energy integrals can be used to study optimal control problems in thick multilevel junctions of type 3:2:2 (see subparagraph "Boundary-value problems" in the Introduction, p. 8).
4. The obtained results may be easily applied to problems in thick m-level junctions of type 3:2:2, where $m \geq 3$. Clearly, the presence of the homogeneous Dirichlet conditions on the surfaces of thin discs at least one of the levels leads to the splitting the starting problem into $m + 1$ independent boundary-value problems as $\varepsilon \to 0$. If the Neumann or Fourier are given on the surfaces of thin discs of all levels, then the corresponding homogenized problem will contain $m + 1$ differential equations that are bound by the conjugation conditions on Ω'; the solution of the homogenized problem will be some m-sheeted function.

Chapter 3
Homogenization of Elliptic Problems in Thick Junctions with Sharp Edges

3.1 Statement of the Problem

In this chapter we suppose that $h_2(r) = h_2(r_0) =: h_2$, $r \in [r_0, r_2]$, where $h_2 \in (0, 1)$ is a fixed number, $h_1 : [r_0, r_1) \mapsto (0, 1)$ is a smooth function in $[r_0, r_1)$, and there exists $\delta_1 > 0$ such that

$$h_1(r) = (r_1 - r)^{1+\gamma} \quad \forall r \in [r_1 - \delta_1, r_1], \tag{3.1}$$

where $\gamma > -1$ is a parameter. Also we suppose that $h_1(s)$ is constant in some neighborhood of r_0, i.e., there exists a positive constant δ_0 such that $h_1(r) = h_1(r_0)$ for all $r \in [r_0, r_0 + \delta_0]$.

According to these assumptions the thick junction Ω_ε, described in Sect. 2.1, looks like in Fig. 3.1 (in case $\gamma > 0$). In accordance with (3.1) geometric structure of the edges of the thin discs $\{\Omega_\varepsilon^{(1)}(j)\}_{j=0}^{N-1}$ significantly depends on γ (see Fig. 3.2).

In Ω_ε, we consider a linear elliptic boundary-value problem

$$\begin{cases} -\Delta_x u_\varepsilon(x) = f_\varepsilon(x), & x \in \Omega_\varepsilon, \\ \partial_\nu u_\varepsilon(x) = 0, & x \in \partial\Omega_\varepsilon^{(1)} \cap \{r > r_0\} \cup \partial\Omega_\varepsilon \cap \{r = r_0\}, \\ \partial_\nu u_\varepsilon(x) + \varepsilon\kappa u_\varepsilon(x) = 0, & x \in \partial\Omega_\varepsilon^{(2)} \cap \{r_0 < r < r_2\}, \\ u_\varepsilon(x) = q_\varepsilon(x), & x \in \partial\Omega_\varepsilon^{(2)} \cap \{r = r_2\}, \\ \partial_{x_2}^p u_\varepsilon|_{x_2=0} = \partial_{x_2}^p u_\varepsilon|_{x_2=l}, & p = 0, 1, \ x \in \partial\Omega_0 \cap \{r < r_0\}. \end{cases} \tag{3.2}$$

Remark 3.1 Hereinafter, we use the precedence (from highest to lowest) for the set operators $\backslash, \cap, \cup, \times$.

In (3.2) κ is a positive constant. Except (2.3), for the right-hand side f_ε in the case $\gamma = 0$ we suppose that there exist positive constants ε_0, C_0 such that

© The Author(s), under exclusive license to Springer Nature Switzerland AG 2019
T. Mel'nyk and D. Sadovyi, *Multiple-Scale Analysis of Boundary-Value Problems in Thick Multi-Level Junctions of Type 3:2:2*, SpringerBriefs in Mathematics, https://doi.org/10.1007/978-3-030-35537-1_3

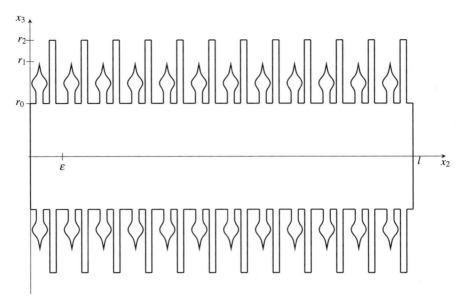

Fig. 3.1 The cross section of the thick junction Ω_ε of type 3:2:2 ($\gamma > 0$, $N = 12$)

$$\int_{\Omega_\varepsilon} F_\varepsilon^2(x)\,dx < C_0 \tag{3.3}$$

for all $\varepsilon \in (0, \varepsilon_0)$, where $F_\varepsilon(x) = \varepsilon^{-1}\left(\overleftrightarrow{f_\varepsilon}(x + \varepsilon\bar{e}_2) - \overleftrightarrow{f_\varepsilon}(x)\right)$, $\bar{e}_2 = (0, 1, 0)$, and $\overleftrightarrow{\varphi}$ is the l-periodic extension of a function $\varphi : \Omega_\varepsilon \to \mathbb{R}$ along the axis Ox_2.

The functions q_ε, $q_0 \in H^{\frac{3}{2}}(\partial\Omega^{(2)} \cap \{r = r_2\})$ and

$$q_\varepsilon \xrightarrow{w} q_0 \quad \text{weakly in } H^{\frac{3}{2}}(\partial\Omega^{(2)} \cap \{r = r_2\}) \text{ as } \varepsilon \to 0. \tag{3.4}$$

Consider a Sobolev space

$$H^1_{\mathrm{per}}(\Omega_\varepsilon) := \{\varphi \in H^1(\Omega_\varepsilon) : \varphi(x_1, 0, x_3) = \varphi(x_1, l, x_3), \; r < r_0\}.$$

Remark 3.2 Here and further we denote by $\varphi(x_1, 0, x_3)$ and $\varphi(x_1, l, x_3)$ the traces of φ on the left and right bases of the cylinder Ω_0.

Definition 3.1 If $\gamma \in (-1, 0]$, then a function $u_\varepsilon \in H^1_{\mathrm{per}}(\Omega_\varepsilon)$ is called a weak solution to problem (3.2) if $u_\varepsilon|_{\partial\Omega_\varepsilon^{(2)} \cap \{r=r_2\}} = q_\varepsilon$ and the integral identity

$$\int_{\Omega_\varepsilon} \nabla_x u_\varepsilon \cdot \nabla_x \varphi \, dx + \varepsilon\kappa \int_{\partial\Omega_\varepsilon^{(2)} \cap \{r_0 < r < r_2\}} u_\varepsilon \varphi \, d\tilde{x} = \int_{\Omega_\varepsilon} f_\varepsilon \varphi \, dx \tag{3.5}$$

holds for all functions $\varphi \in H^1_{\mathrm{per}}(\Omega_\varepsilon)$ such that $\varphi|_{\partial\Omega_\varepsilon^{(2)} \cap \{r=r_2\}} = 0$.

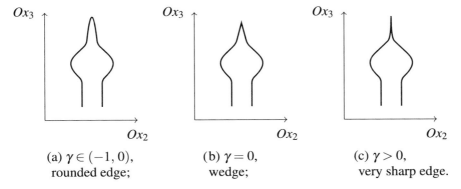

(a) $\gamma \in (-1, 0)$,
rounded edge;

(b) $\gamma = 0$,
wedge;

(c) $\gamma > 0$,
very sharp edge.

Fig. 3.2 Cross sections of the edges of the thin discs from the first level

Let

$$R_\alpha^\beta = \{x \in \mathbb{R}^3 : r \in (\alpha, \beta), \ x_2 \in (0, l)\},$$

where $0 \leq \alpha < \beta$. We denote by $C_{\mathrm{per}}^\infty(\overline{\Omega}_\varepsilon, \partial\Omega_\varepsilon^{(1)} \cap \{r = r_1\})$ the set of smooth functions $\varphi : \overline{\Omega}_\varepsilon \mapsto \mathbb{R}$ vanishing in some subset $\Omega_\varepsilon^{(1)} \cap R_{r_1-\alpha}^{r_1}$ (not necessarily the same for all functions) that are l-periodic in x_2. The closure of $C_{\mathrm{per}}^\infty(\overline{\Omega}_\varepsilon, \partial\Omega_\varepsilon^{(1)} \cap \{r = r_1\})$ in $H^1(\Omega_\varepsilon)$ denotes by $H_{\mathrm{per}}^1(\Omega_\varepsilon, \partial\Omega_\varepsilon^{(1)} \cap \{r = r_1\})$.

Remark 3.3 In case $\gamma = 0$, we conclude that $H_{\mathrm{per}}^1(\Omega_\varepsilon, \partial\Omega_\varepsilon^{(1)} \cap \{r = r_1\}) = H_{\mathrm{per}}^1(\Omega_\varepsilon)$ (see Corollary 3.1). But if $\gamma > 0$ then the boundary $\partial\Omega_\varepsilon$ is not Lipschitz, and we cannot state that these spaces coincide (see [67]).

Definition 3.2 In case $\gamma > 0$, we say that a function $u_\varepsilon \in H_{\mathrm{per}}^1(\Omega_\varepsilon, \partial\Omega_\varepsilon^{(1)} \cap \{r = r_1\})$ is a weak solution to problem (3.2) if $u_\varepsilon|_{\partial\Omega_\varepsilon^{(2)} \cap \{r=r_2\}} = q_\varepsilon$ and integral identity (3.5) holds for all $\varphi \in H_{\mathrm{per}}^1(\Omega_\varepsilon, \partial\Omega_\varepsilon^{(1)} \cap \{r = r_1\})$ such that $\varphi|_{\partial\Omega_\varepsilon^{(2)} \cap \{r=r_2\}} = 0$.

3.2 Auxiliary Statements

3.2.1 Cases of Rounded and Linear Edges

In these cases, the boundary of Ω_ε is Lipschitz (see Fig. 3.2a, b). Clearly, for a.e. $x \in \partial\Omega_\varepsilon^{(1)} \cap \{r > r_0\}$ the outward unit normal at x is defined by (2.8) and the integral identity (2.9) holds. For thin discs from the second level the identity (2.9) has the form

$$\frac{\varepsilon h_2}{2} \int_{\partial\Omega_\varepsilon^{(2)} \cap \{r_0 < r < r_2\}} \varphi \, d\tilde{x} = \int_{\Omega_\varepsilon^{(2)}} \varphi \, dx - \varepsilon \int_{\Omega_\varepsilon^{(2)}} Y_2\left(\frac{x_2}{\varepsilon}\right) \partial_{x_2}\varphi \, dx \quad \forall \varphi \in H^1(\Omega_\varepsilon).$$

$$(3.6)$$

Lemma 3.1 *For $\gamma \in (-1, 0]$, there exist positive constants ε_0, C_0 such that for all $\varepsilon \in (0, \varepsilon_0)$ and $\varphi \in H^1(\Omega_\varepsilon)$ the estimate (2.10) holds.*

Proof Let $\varphi \in C^\infty(\overline{\Omega_\varepsilon})$. We choose a fixed $\alpha \in (0, \frac{\delta_1}{2})$ such that

$$\min_{r \in [r_0, r_1 - \alpha]} h_1(r) = h_1(r_1 - \alpha) = \alpha^{1+\gamma}. \tag{3.7}$$

Similarly as in [32], we can prove the following identity:

$$\frac{\varepsilon}{2} \int_{\partial\Omega_\varepsilon^{(1)} \cap \{r > r_0\} \cap R_{r_0}^{r_1 - \alpha}} \frac{h_1(r)\psi}{N_\varepsilon^{(1)}(r)} \, d\sigma_x = \int_{\Omega_\varepsilon^{(1)} \cap R_{r_0}^{r_1 - \alpha}} \psi \, dx - \varepsilon \int_{\Omega_\varepsilon^{(1)} \cap R_{r_0}^{r_1 - \alpha}} Y_1\left(\frac{x_2}{\varepsilon}\right) \partial_{x_2}\psi \, dx$$

for all $\psi \in H^1(\Omega_\varepsilon)$. Taking into account (3.7), the last identity, and the facts that $\max_{\mathbb{R}} |Y_1| \le 2$ and $\max_{[r_0, r_1 - \alpha]} |h_1'| \le c_0\alpha^\gamma$, we derive

$$\|\varphi\|_{L^2(\partial\Omega_\varepsilon^{(1)} \cap \{r > r_0\} \cap R_{r_0}^{r_1 - \alpha})} \le c_1\alpha^{-\frac{1}{2}}\varepsilon^{-\frac{1}{2}}\|\varphi\|_{H^1(\Omega_\varepsilon^{(1)} \cap R_{r_0}^{r_1 - \alpha})}, \quad \forall \varphi \in H^1(\Omega_\varepsilon). \tag{3.8}$$

Let (r, x_2, θ) be the cylindric coordinates of $x \in \partial\Omega_\varepsilon^{(1)}(j) \cap \{r > r_0\} \cap R_{r_1 - \alpha}^{r_1}$. Using (3.1) we get

$$r = r_1 - \left|\frac{2(x_2 - \varepsilon(j + l_1))}{\varepsilon}\right|^{\frac{1}{1+\gamma}} =: \rho_\varepsilon(x_2), \quad x_2 \in I_\varepsilon^{(1)}(j, h_1(r_1 - \alpha)).$$

Consider the following obvious inequality:

$$\varphi^2|_{r=\rho_\varepsilon(x_2)} \le 2\alpha \int_{r_1 - 2\alpha}^{\rho_\varepsilon(x_2)} (\partial_r\varphi)^2 \, dr + 2\varphi^2(x), \quad x \in \Omega_\varepsilon^{(1)}(j) \cap R_{r_1 - \alpha}^{r_1}.$$

Multiplying this inequality by the surface element of $\left(\partial\Omega_\varepsilon^{(1)}(j) \cap \{r > r_0\} \cap R_{r_1 - \alpha}^{r_1}\right)$ and integrating it over $r \in (r_1 - 2\alpha, \rho_\varepsilon(x_2))$, $x_2 \in I_\varepsilon^{(1)}(j, h_1(r_1 - \alpha))$, $\theta \in [0, 2\pi)$, we deduce

$$\|\varphi\|_{L^2(\partial\Omega_\varepsilon^{(1)}(j) \cap \{r_0 < r < r_1\} \cap R_{r_1 - \alpha}^{r_1})} \le c_2\varepsilon^{-\frac{1}{2}}\alpha^{-\frac{1+\gamma}{2}}\|\varphi\|_{H^1(\Omega_\varepsilon^{(1)}(j) \cap R_{r_1 - 2\alpha}^{r_1})} \tag{3.9}$$

for all $j = 0, \ldots, N - 1$. Inequalities (3.9) together with (3.8) provide the desired inequality. By the closure the estimate (2.10) remains valid for all $\varphi \in H^1(\Omega_\varepsilon)$. $\quad\square$

Similar as in [121, Lemma 4.1] we prove the following statement.

Lemma 3.2 *In case $\gamma \in (-1, 0)$ there exist positive constants ε_0, C_0 such that for all $\varepsilon \in (0, \varepsilon_0)$ and $\varphi \in H^1(\Omega_\varepsilon)$ the following inequality holds:*

$$\|h_1^{-1}\varphi\|_{L^2(\Omega_\varepsilon^{(1)})} \le C_1\|\varphi\|_{H^1(\Omega_\varepsilon^{(1)})}.$$

3.2.2   Case of Very Sharp Edges

In this case, the boundary of Ω_ε is not Lipschitz (see Fig. 3.2c). First we prove some properties of functions from the space $H^1_{\mathrm{per}}(\Omega_\varepsilon, \partial\Omega^{(1)}_\varepsilon \cap \{r = r_1\})$ for $\gamma \geq 0$.

Lemma 3.3 *Every function $\varphi \in H^1_{\mathrm{per}}(\Omega_\varepsilon)$ such that*

(1) $|\varphi| \leq C_2 |\ln(r_1 - r)|^{1/2}$ a.e. in some subset $\Omega^{(1)}_\varepsilon \cap R^{r_1}_{r_1 - \delta}$ for $\gamma = 0$,

or

(2) $|\varphi| \leq C_2 (r_1 - r)^{-\gamma/2}$ a.e. in some subset $\Omega^{(1)}_\varepsilon \cap R^{r_1}_{r_1 - \delta}$ for $\gamma > 0$,

belongs to $H^1_{\mathrm{per}}(\Omega_\varepsilon, \partial\Omega^{(1)}_\varepsilon \cap \{r = r_1\})$. Here δ is some positive number.

Proof For the proof we will use the approach of [133, Theorem 1]. Let φ be an arbitrary function that satisfies the Lemma's assumptions. Consider a cutoff function

$$\chi_\alpha(r) = \begin{cases} 1, & r \leq r_1 - \alpha_1, \\ (\ln|\ln\alpha|)^\eta - (\ln|\ln(r_1 - r)|)^\eta, & r_1 - \alpha_1 < r \leq r_1 - \alpha, \\ 0, & r_1 - \alpha < r < r_1, \end{cases}$$

where $\eta \in (0, 1/2)$, α_1 is defined from the equality

$$(\ln|\ln\alpha|)^\eta - (\ln|\ln\alpha_1|)^\eta = 1,$$

and α is some positive number and it is so small that $\alpha_1 \in (0, \delta)$. It is easy to verify that $\chi_\alpha \varphi \in H^1_{\mathrm{per}}(\Omega_\varepsilon, \partial\Omega^{(1)}_\varepsilon \cap \{r = r_1\})$. Then we have

$$\|\varphi - \chi_\alpha\varphi\|^2_{H^1(\Omega_\varepsilon)} = \|\varphi(1 - \chi_\alpha)\|^2_{H^1(\Omega_\varepsilon)} = \int_{\Omega^{(1)}_\varepsilon} \varphi^2(1 - \chi_\alpha)^2 \, dx$$

$$+ \int_{\Omega^{(1)}_\varepsilon} |\nabla_x\varphi|^2(1 - \chi_\alpha)^2 \, dx + 2\int_{\Omega^{(1)}_\varepsilon} \nabla_x\varphi \cdot \nabla_x(1 - \chi_\alpha)\varphi(1 - \chi_\alpha) \, dx$$

$$+ \int_{\Omega^{(1)}_\varepsilon} \varphi^2|\nabla_x(1 - \chi_\alpha)|^2 \, dx =: I_{\varepsilon,1}(\alpha, \varphi) + I_{\varepsilon,2}(\alpha, \varphi) + 2I_{\varepsilon,3}(\alpha, \varphi) + I_{\varepsilon,4}(\alpha, \varphi).$$

It is easy to see that $I_{\varepsilon,1}(\alpha, \varphi) \to 0$ and $I_{\varepsilon,2}(\alpha, \varphi) \to 0$ as $\alpha \to 0$ since $(1 - \chi_\alpha)^2 \leq 1$, $\lim_{\alpha \to 0}(1 - \chi_\alpha)^2 = 0$, and $\|\varphi\|_{H^1(\Omega_\varepsilon)} < +\infty$.

If $\gamma = 0$, then with the help of the first condition of the lemma and (3.1) we derive

$$I_{\varepsilon,4}(\alpha, \varphi) = \int_{\Omega^{(1)}_\varepsilon \cap R^{r_1-\alpha}_{r_1-\alpha_1}} \varphi^2|\nabla_x(1 - \chi_\alpha)|^2 \, dx$$

$$\leq c_0 \int_{\Omega^{(1)}_\varepsilon \cap R^{r_1-\alpha}_{r_1-\alpha_1}} \frac{|\ln(r_1 - r)| \, dx}{(r_1 - r)^2 |\ln(r_1 - r)|^2 (\ln|\ln(r_1 - r)|)^{2-2\eta}}$$

$$\leq c_1 \int_{r_1-\alpha_1}^{r_1-\alpha} \frac{r \, dr}{(r_1 - r)|\ln(r_1 - r)|(\ln|\ln(r_1 - r)|)^{2-2\eta}} \longrightarrow 0 \quad \text{as } \alpha \to 0.$$

If $\gamma > 0$, then by the same way we get

$$I_{\varepsilon,4}(\alpha, \varphi) = \int_{\Omega_\varepsilon^{(1)} \cap R_{r_1-\alpha_1}^{r_1-\alpha}} \varphi^2 |\nabla_x (1 - \chi_\alpha)|^2 \, dx$$

$$\leq c_2 \int_{\Omega_\varepsilon^{(1)} \cap R_{r_1-\alpha_1}^{r_1-\alpha}} \frac{(r_1 - r)^{-\gamma} \, dx}{(r_1 - r)^2 |\ln(r_1 - r)|^2 (\ln |\ln(r_1 - r)|)^{2-2\eta}}$$

$$\leq c_3 \int_{r_1-\alpha_1}^{r_1-\alpha} \frac{r \, dr}{(r_1 - r)|\ln(r_1 - r)|^2} \longrightarrow 0 \quad \text{as } \alpha \to 0.$$

Since $|I_{\varepsilon,3}(\alpha, \varphi)| \leq \sqrt{I_{\varepsilon,2}(\alpha, \varphi) I_{\varepsilon,4}(\alpha, \varphi)}$, $\lim_{\alpha \to 0} |I_{\varepsilon,3}(\alpha, \varphi)| = 0$.

Thus, $\chi_\alpha \varphi \rightarrow \varphi$ strongly in $H^1(\Omega_\varepsilon)$ as $\alpha \to 0$ and consequently $\varphi \in H^1_{\text{per}}(\Omega_\varepsilon, \partial\Omega_\varepsilon^{(1)} \cap \{r = r_1\})$.

Corollary 3.1 *If $\gamma \geq 0$, then any function $\varphi \in C^\infty(\overline{\Omega_\varepsilon})$ such that $\varphi(x_1, 0, x_3) = \varphi(x_1, l, x_3)$ can be approximated with a sequence from $C^\infty_{\text{per}}(\overline{\Omega_\varepsilon}, \partial\Omega_\varepsilon^{(1)} \cap \{r = r_1\})$ in the standard norm of $H^1(\Omega_\varepsilon)$.*

3.2.3 A Priori Estimate

Similarly as in [27], we can prove that the norms $\| \cdot \|_{H^1(\Omega_\varepsilon)}$ and

$$\|\varphi\|_\varepsilon := \left(\int_{\Omega_\varepsilon} |\nabla_x \varphi|^2 \, dx + \varepsilon\kappa \int_{\partial\Omega_\varepsilon^{(2)} \cap \{r_0 < r < r_2\}} \varphi^2 \, d\tilde{x} \right)^{\frac{1}{2}}$$

are uniformly in ε equivalent in $H^1_{\text{per}}(\Omega_\varepsilon)$ (in $H^1_{\text{per}}(\Omega_\varepsilon, \partial\Omega_\varepsilon^{(1)} \cap \{r = r_1\})$) for all $\gamma \in (-1, 0]$ ($\gamma > 0$), i.e., there exist positive constants ε_0, C_3, C_4 such that $C_3\|\varphi\|_{H^1(\Omega_\varepsilon)} \leq \|\varphi\|_\varepsilon \leq C_4\|\varphi\|_{H^1(\Omega_\varepsilon)}$ for all $\varepsilon \in (0, \varepsilon_0)$ and $\varphi \in H^1_{\text{per}}(\Omega_\varepsilon)$ ($\varphi \in H^1_{\text{per}}(\Omega_\varepsilon, \partial\Omega_\varepsilon^{(1)} \cap \{r = r_1\})$).

Since $q_\varepsilon \in H^{\frac{3}{2}}(\partial\Omega_\varepsilon^{(2)} \cap \{r = r_2\})$, there exists a function $q_\varepsilon \in H^1(\Omega_\varepsilon)$ such that $\|q_\varepsilon\|_{H^1(\Omega_\varepsilon)} \leq c_0\|q_\varepsilon\|_{H^{\frac{1}{2}}(\partial\Omega_\varepsilon^{(2)} \cap \{r=r_2\})}$ and $\text{supp}\, q_\varepsilon \subset \Omega_\varepsilon^{(2)} \cap R_{r_0+\delta}^{r_2}$ for some $\delta > 0$.

Then the function $v_\varepsilon = u_\varepsilon - q_\varepsilon$ satisfies the integral identity

$$\int_{\Omega_\varepsilon} \nabla_x v_\varepsilon \cdot \nabla_x \varphi \, dx + \varepsilon\kappa \int_{\partial\Omega_\varepsilon^{(2)} \cap \{r_0 < r < r_2\}} v_\varepsilon \varphi \, d\tilde{x}$$

$$= \int_{\Omega_\varepsilon} f_\varepsilon \varphi \, dx - \int_{\Omega_\varepsilon^{(2)}} \nabla_x q_\varepsilon \cdot \nabla_x \varphi \, dx - \varepsilon\kappa \int_{\partial\Omega_\varepsilon^{(2)} \cap \{r_0 < r < r_2\}} q_\varepsilon \varphi \, d\tilde{x},$$

where φ is arbitrary function from $H^1_{\text{per}}(\Omega_\varepsilon)$ such that $\varphi|_{\partial\Omega_\varepsilon^{(2)} \cap \{r=r_2\}} = 0$ if $\gamma \in (-1, 0]$, or $\varphi \in H^1_{\text{per}}(\Omega_\varepsilon, \partial\Omega_\varepsilon^{(1)} \cap \{r = r_1\})$ if $\gamma > 0$. According to (2.3), (3.4), and (2.10), the right-hand side of the last identity is a linear continuous functional

$$\mathsf{L}_\varepsilon(\varphi) := \int_{\Omega_\varepsilon} f_\varepsilon \varphi \, dx - \int_{\Omega_\varepsilon^{(2)}} \nabla_x q_\varepsilon \cdot \nabla_x \varphi \, dx - \varepsilon \kappa \int_{\partial \Omega_\varepsilon^{(2)} \cap \{r_0 < r < r_2\}} q_\varepsilon \varphi \, d\tilde{x}$$

on $H^1_{\mathrm{per}}(\Omega_\varepsilon)$ if $\gamma \in (-1, 0]$, or on $H^1_{\mathrm{per}}(\Omega_\varepsilon, \partial \Omega_\varepsilon^{(1)} \cap \{r = r_1\})$ if $\gamma > 0$. Since the norm of the functional L_ε is uniformly bounded in ε, then due to the Riesz representation theorem we have the following statement.

Lemma 3.4 *For all $\gamma > -1$ and $\varepsilon > 0$, there exists a unique weak solution u_ε to problem (3.2). In addition, there exist positive constants ε_0, C_5 such that for all $\varepsilon \in (0, \varepsilon_0)$ the following inequality holds:*

$$\|u_\varepsilon\|_{H^1(\Omega_\varepsilon)} \le C_1 \left(\|f_\varepsilon\|_{L^2(\Omega_\varepsilon)} + \|q_\varepsilon\|_{H^{\frac{1}{2}}(\partial \Omega_\varepsilon^{(2)} \cap \{r=r_2\})} \right).$$

3.2.4 Weighted Sobolev Spaces

Let $\widetilde{H}_{h_1}(\Omega^{(1)})$ be the closure of $C^\infty(\overline{\Omega^{(1)}})$ with the norm $\| \cdot \|_{h_1}$, where

$$\|\varphi\|_{h_1} = \sqrt{\int_{\Omega^{(1)}} (h_1(r)|\nabla_{\tilde{x}}\varphi|^2 + \varphi^2) \, dx}.$$

Now consider the space $C^\infty(\overline{\Omega^{(1)}}, \partial \Omega^{(1)} \cap \{r = r_1\})$ of smooth functions that vanish in some subset $\Omega^{(1)} \cap R^{r_1}_{r_1 - \alpha}$. We introduce the space

$$\widetilde{H}_{h_1}(\Omega^{(1)}, \partial \Omega^{(1)} \cap \{r = r_1\}) := \left(C^\infty(\overline{\Omega^{(1)}}, \partial \Omega^{(1)} \cap \{r = r_1\}), \ \| \cdot \|_{h_1} \right),$$

i.e., it is the closure of $C^\infty(\overline{\Omega^{(1)}}, \partial \Omega^{(1)} \cap \{r = r_1\})$ with the weighted norm $\| \cdot \|_{h_1}$.

According to (3.1), we have $H^1(\Omega^{(1)}) \subset \widetilde{H}_{h_1}(\Omega^{(1)})$ for all $\gamma > -1$, and this embedding is continuous.

The following statement shows that functions from $\widetilde{H}_{h_1}(\Omega^{(1)}, \partial \Omega^{(1)} \cap \{r = r_1\})$ can tend to infinity as $r \to r_1$ in the case $\gamma \ge 0$.

Lemma 3.5 *A function $\varphi \in \widetilde{H}_{h_1}(\Omega^{(1)})$ such that*

(1) $|\varphi| \le C_6 |\ln(r_1 - r)|^{\frac{1}{2}}$ a.e. in some domain $\Omega^{(1)} \cap R^{r_1}_{r_1 - \delta}$ in the case $\gamma = 0$,

or

(2) $|\varphi| \le C_6 (r_1 - r)^{-\frac{\gamma}{2}}$ a.e. in some domain $\Omega^{(1)} \cap R^{r_1}_{r_1 - \delta}$ in the case $\gamma > 0$,

belongs to the weighted Sobolev space $\widetilde{H}_{h_1}(\Omega^{(1)}, \partial \Omega^{(1)} \cap \{r = r_1\})$.

The proof of this statement is similar to the proof of the second statement in [133, Theorem 1]. This statement implies that any function from $C^\infty(\overline{\Omega^{(1)}})$ can be approximated by functions from $C^\infty(\overline{\Omega^{(1)}}, \partial \Omega^{(1)} \cap \{r = r_1\})$ in the norm $\| \cdot \|_{h_1}$.

3.3 Special Multivalued Extension

In this section we construct a multivalued extension of the weak solution u_ε, which preserves H^1-smoothness of the solution in the case $\gamma = 0$. First we extend the solution from the thin discs $\Omega_\varepsilon^{(1)}$ into the domain $\Omega^{(1)}$.

Theorem 3.1 *If $\gamma = 0$ and conditions (2.3), (3.3), and (3.4) hold, then there exist positive constants ε_0, C_0, and an extension $E_{u_\varepsilon}^{(1)} \in H^1(\Omega_0 \cup \Omega^{(1)})$ of the weak solution u_ε to problem (3.2) such that for all $\varepsilon \in (0, \varepsilon_0)$ the following inequality holds:*

$$\|E_{u_\varepsilon}^{(1)}\|_{H^1(\Omega^{(1)})} \le C_0. \tag{3.10}$$

Proof **1.** Similarly as in [32], we can show that

$$\|U_\varepsilon\|_{H^1(\Omega_\varepsilon)} \le c_0 \left(\|F_\varepsilon\|_{L^2(\Omega_\varepsilon)} + \|q_\varepsilon\|_{H^{\frac{3}{2}}(\partial\Omega_\varepsilon^{(2)} \cap \{r=r_2\})} \right), \tag{3.11}$$

where $U_\varepsilon(x) = \varepsilon^{-1}\left(\overleftrightarrow{u_\varepsilon}(x + \varepsilon\bar{e}_2) - \overleftrightarrow{u_\varepsilon}(x) \right)$.
   2. Let

$$\widetilde{\Omega_\varepsilon^{(i)}}(j) := \{x \in \mathbb{R}^3 : \varepsilon h_i(r)/2 \le x_2 - \varepsilon(j + l_i) \le \varepsilon(1 - h_i(r)/2)\}.$$

This set is situated between two neighboring thin discs $\Omega_\varepsilon^{(i)}(j)$ and $\Omega_\varepsilon^{(i)}(j+1)$.
   At first, we construct the extension $E_{u_\varepsilon}^{(1)}$ of u_ε into $R_{r_0+\varepsilon}^{r_1}$ with the formula

$$E_{u_\varepsilon}^{(1)}(x) = a_\varepsilon(j, \tilde{x}) + b_\varepsilon(j, \tilde{x})\left(x_2 - \varepsilon(j + l_1 + h_1(r)/2)\right),$$

for all $x \in \widetilde{\Omega_\varepsilon^{(i)}}(j) \cap R_{r_0+\varepsilon}^{r_1}$ and $j = -1, \ldots, N-1$, where

$$a_\varepsilon(j, \tilde{x}) = \overleftrightarrow{u_\varepsilon}(x_1, \varepsilon(j + l_1 + h_1(r)/2), x_3),$$

$$b_\varepsilon(j, \tilde{x}) = \frac{1}{\varepsilon(1 - h_1(r))} \left(\overleftrightarrow{u_\varepsilon}(x_1, \varepsilon(j + 1 + l_1 - h_1(r)/2), x_3) - a_\varepsilon(j, \tilde{x}) \right).$$

We choose α as in the proof of Lemma 3.1 (see (3.7)). Similarly to [32], we can show that

$$\|E_{u_\varepsilon}^{(1)}\|_{H^1(R_{r_0+\varepsilon}^{r_1-\alpha})}^2 \le c_1 \Big(\alpha^{-1}(\|U_\varepsilon\|_{H^1(\Omega_\varepsilon^{(1)})}^2 + \|u_\varepsilon\|_{H^1(\Omega_\varepsilon^{(1)})}^2)$$
$$+ \|F_\varepsilon\|_{L^2(\Omega_\varepsilon^{(1)})}^2 + \|f_\varepsilon\|_{L^2(\Omega_\varepsilon^{(1)})}^2 \Big). \tag{3.12}$$

3. Direct calculations give

$$\|E_{u_\varepsilon}^{(1)}\|_{H^1(\widetilde{\Omega_\varepsilon^{(i)}}(j) \cap R_{r_1-\alpha}^{r_1})}^2 \le c_2 \int_{R_{r_1-\alpha}^{r_1}} \left(\varepsilon(a_\varepsilon^2 + |\nabla_{\tilde{x}} a_\varepsilon|^2) + \varepsilon^3(b_\varepsilon^2 + |\nabla_{\tilde{x}} b_\varepsilon|^2) + \varepsilon b_\varepsilon^2 \right) dx.$$

Utilizing (3.9), we prove that

$$\varepsilon \int_{R^{r_1}_{r_1-\alpha}} a_\varepsilon^2 \, dx = \varepsilon \int_{R^{r_1}_{r_1-\alpha}} u_\varepsilon^2|_{x_2=\varepsilon(h_1(r)/2+j+l_1)} \, dx$$

$$\leq c_3\varepsilon \|(N_\varepsilon^{(1)}(r))^{-\frac{1}{2}} u_\varepsilon\|^2_{L^2(\Omega_\varepsilon^{(1)}(j)\cap R^{r_1}_{r_1-\alpha})} \leq c_4 \|u_\varepsilon\|^2_{H^1(\Omega_\varepsilon^{(1)}(j)\cap R^{r_1}_{r_1-2\alpha})}.$$

With the help of the inequality

$$(v(s)-v(0))^2 \leq s\int_0^s (v'(t))^2 \, dt \quad \text{for all } v \in H^1((0,s))$$

and (3.9) we derive that

$$\varepsilon \int_{R^{r_1}_{r_1-\alpha}} b_\varepsilon^2 \, dx \leq c_5\left(\|U_\varepsilon\|^2_{H^1(\Omega_\varepsilon^{(1)}(j)\cap R^{r_1}_{r_1-2\alpha})} + \|u_\varepsilon\|^2_{H^1(\Omega_\varepsilon^{(1)}(j)\cap R^{r_1}_{r_1-\alpha})}\right),$$

$$\varepsilon \int_{R^{r_1}_{r_1-\alpha}} |\nabla_{\tilde{x}} a_\varepsilon|^2 \, dx \leq c_6 \|u_\varepsilon\|^2_{H^2(\Omega_\varepsilon^{(1)}(j)\cap R^{r_1}_{r_1-2\alpha})},$$

$$\varepsilon^3 \int_{R^{r_1}_{r_1-\alpha}} |\nabla_{\tilde{x}} b_\varepsilon|^2 \, dx \leq c_7\varepsilon^2\left(\|U_\varepsilon\|^2_{H^2(\Omega_\varepsilon^{(1)}(j)\cap R^{r_1}_{r_1-2\alpha})} + \|u_\varepsilon\|^2_{H^2(\Omega_\varepsilon^{(1)}(j)\cap R^{r_1}_{r_1-2\alpha})}\right).$$

Consider the smooth cutoff function

$$\chi(r) = \begin{cases} 0, & r \in [0, r_0 + \delta_0/3], \\ 1, & r \in [r_0 + 2\delta_0/3, r_1]. \end{cases}$$

Taking into account the properties of the function h_1 and so-called the second energy inequality (see [52, Sect. 8.2]), we conclude that

$$\|u_\varepsilon\|_{H^2(\Omega_\varepsilon^{(1)}(j)\cap R^{r_1}_{r_1-2\alpha})} \leq \|\chi u_\varepsilon\|_{H^2(\Omega_\varepsilon^{(1)}(j)\cap R^{r_1}_{r_0+\delta_0/3})}$$

$$\leq c_8\left(\|u_\varepsilon\|_{H^1(\Omega_\varepsilon^{(1)}(j))} + \|f_\varepsilon\|_{L^2(\Omega_\varepsilon^{(1)}(j))}\right)$$

and

$$\|U_\varepsilon\|_{H^2(\Omega_\varepsilon^{(1)}(j)\cap R^{r_1}_{r_1-2\alpha})} \leq c_9\left(\|U_\varepsilon\|_{H^1(\Omega_\varepsilon^{(1)}(j))} + \|F_\varepsilon\|_{L^2(\Omega_\varepsilon^{(1)}(j))}\right).$$

Using those inequalities, we derive the estimate

$$\|E^{(1)}_{u_\varepsilon}\|^2_{H^1(R^{r_1}_{r_1-\alpha})} \leq c_{10}\left(\|u_\varepsilon\|^2_{H^1(\Omega_\varepsilon^{(1)})} + \|U_\varepsilon\|^2_{H^1(\Omega_\varepsilon^{(1)})} + \|f_\varepsilon\|^2_{L^2(\Omega_\varepsilon^{(1)})} + \|F_\varepsilon\|^2_{L^2(\Omega_\varepsilon^{(1)})}\right).$$

$$(3.13)$$

4. Now it remains to extend $E_{u_\varepsilon}^{(1)}$ into $\widetilde{\Omega_\varepsilon^{(1)}}(j) \cap R_{r_0}^{r_0+\varepsilon}$, where $j = -1, \ldots, N - 1$. Similarly as in the third part of the proof of Theorem 2 in [31], we get

$$\|E_{u_\varepsilon}^{(1)}\|_{H^1(\Omega^{(1)})} \le c_1 1 (\|E_{u_\varepsilon}^{(1)}\|_{H^1(\Omega_\varepsilon^{(1)} \cup R_{r_0+\varepsilon}^{r_1})} + \|u_\varepsilon\|_{H^1(\Omega_0)}). \tag{3.14}$$

Thus, according to (2.3), (3.3), (3.4), Lemma 3.4, (3.11), (3.12), (3.13), and (3.14) the extension $E_{u_\varepsilon}^{(1)}(x)$, $x \in \Omega^{(1)}$, satisfies the estimate of Theorem 3.1.

Next we construct the extension of u_ε from the discs from the second level into $\Omega^{(2)}$.

Theorem 3.2 *If $\gamma = 0$ and conditions (2.3), (3.3), and (3.4) hold, then there exist positive constants ε_0, C_1 and an extension $E_{u_\varepsilon}^{(2)} \in H^1(\Omega_0 \cup \Omega^{(2)})$ of the weak solution u_ε to problem (3.2) such that for all $\varepsilon \in (0, \varepsilon_0)$: $E_{u_\varepsilon}^{(2)}|_{\partial\Omega^{(2)} \cap \{r=r_2\}} = q_\varepsilon$ and*

$$\|E_{u_\varepsilon}^{(2)}\|_{H^1(\Omega^{(2)})} \le C_1.$$

Proof By the same formulas as in the second part of the proof of Theorem 3.1, we construct the extension $E_{u_\varepsilon}^{(2)}$ of u_ε into the domain $R_{r_0+\varepsilon}^{r_2-\varepsilon}$ such that

$$\|E_{u_\varepsilon}^{(2)}\|_{H^1(R_{r_0+\varepsilon}^{r_2-\varepsilon})}^2$$
$$\le c_1 2 \left(\|u_\varepsilon\|_{H^1(\Omega_\varepsilon^{(2)})}^2 + \|U_\varepsilon\|_{H^1(\Omega_\varepsilon^{(2)})}^2 + \|f_\varepsilon\|_{L^2(\Omega_\varepsilon^{(2)})}^2 + \|F_\varepsilon\|_{L^2(\Omega_\varepsilon^{(2)})}^2 \right) \le c_1 3.$$

Now we extend u_ε into the outside of $\Omega^{(2)}$. Since $q_\varepsilon \in H^{\frac{1}{2}}(\partial\Omega^{(2)} \cap \{r = r_2\})$, then there exist a constant $a > 0$ and a function $\widetilde{q}_\varepsilon \in H^1(R_{r_2}^{r_2+a})$ such that

$$\widetilde{q}_\varepsilon|_{\partial\Omega^{(2)} \cap \{r=r_2\}} = q_\varepsilon \quad \text{and} \quad \|\widetilde{q}_\varepsilon\|_{H^1(R_{r_2}^{r_2+a})} \le c_1 4 \|q_\varepsilon\|_{H^{\frac{1}{2}}(\partial\Omega^{(2)} \cap \{r=r_2\})}.$$

We henceforth regard that $E_{u_\varepsilon}^{(2)}(x) = \widetilde{q}_\varepsilon(x)$ for $x \in R_{r_2}^{r_2+a}$. Obviously, $E_{u_\varepsilon}^{(2)} \in H^1$ $(\Omega_\varepsilon^{(2)} \cup R_{r_0+\varepsilon}^{r_2-\varepsilon} \cup R_{r_2}^{r_2+a})$ and

$$\|E_{u_\varepsilon}^{(2)}\|_{H^1(R_{r_2}^{r_2+a})} = \|\widetilde{q}_\varepsilon\|_{H^1(R_{r_2}^{r_2+a})} \le c_1 5 \|q_\varepsilon\|_{H^{\frac{1}{2}}(\partial\Omega^{(2)} \cap \{r=r_2\})}.$$

Now it remains to extend $E_{u_\varepsilon}^{(2)}$ into $\widetilde{\Omega_\varepsilon^{(2)}}(j) \cap R_{r_0}^{r_0+\varepsilon}$ and $\widetilde{\Omega_\varepsilon^{(2)}}(j) \cap R_{r_2-\varepsilon}^{r_2}$, $j = -1, \ldots, N - 1$. Again we do this similarly as in the third part of the proof of Theorem 2 in [31] and get

$$\|E_{u_\varepsilon}^{(2)}\|_{H^1(\Omega_0 \cup \Omega^{(2)})} \le c_1 6 \left(\|E_{u_\varepsilon}^{(2)}\|_{H^1(\Omega_\varepsilon^{(2)} \cup R_{r_0+\varepsilon}^{r_2-\varepsilon} \cup R_{r_2}^{r_2+a})} + \|u_\varepsilon\|_{H^1(\Omega_0)} \right).$$

Thus, the desired extension is the restriction of $E_{u_\varepsilon}^{(2)}|_{\Omega_0 \cup \Omega^{(2)}}$.

With the help of the constructed extensions $E_{u_\varepsilon}^{(1)}$ and $E_{u_\varepsilon}^{(2)}$ we define a multivalued extension

$$\mathbf{E}_{u_\varepsilon} := \left(u_\varepsilon|_{\Omega_0}, \ E_{u_\varepsilon}^{(1)}|_{\Omega^{(1)}}, \ E_{u_\varepsilon}^{(2)}|_{\Omega^{(2)}} \right) \tag{3.15}$$

of the solution u_ε to problem (3.2), which belongs to the anisotropic Sobolev space of multivalued functions

$$\mathbf{H}_{\mathrm{per}}^1 = \left\{ \mathbf{p} = (p_0, \ p_1, \ p_2) : \ p_0 \in H_{\mathrm{per}}^1(\Omega_0), \ \ p_i \in H^1(\Omega^{(i)}), \ \ i = 1, \ 2, \right.$$

$$\left. p_0|_{\Omega'} = p_1|_{\Omega'} = p_2|_{\Omega'} \right\}$$

with the scalar product

$$(\mathbf{p}, \ \mathbf{q})_{\mathbf{H}_{\mathrm{per}}^1} = \int_{\Omega_0} (\nabla_x p_0 \cdot \nabla_x q_0 + p_0 q_0) \, dx + \sum_{i=1}^2 \int_{\Omega^{(i)}} (\nabla_x p_i \cdot \nabla_x q_i + p_i q_i) \, dx,$$

where $\mathbf{p} = (p_0, \ p_1, \ p_2)$, $\mathbf{q} = (q_0, \ q_1, \ q_2) \in \mathbf{H}_{\mathrm{per}}^1$.

Lemma 3.4, Theorems 3.1, and 3.2 imply that in the case $\gamma = 0$ there exist positive constants ε_0, C_2 such that for all $\varepsilon \in (0, \ \varepsilon_0)$ the multivalued extension $\mathbf{E}_{u_\varepsilon}$ of the solution u_ε to problem (3.2) satisfies the inequality

$$\|\mathbf{E}_{u_\varepsilon}\|_{\mathbf{H}_{\mathrm{per}}^1} \leq C_2, \tag{3.16}$$

where $\| \cdot \|_{\mathbf{H}_{\mathrm{per}}^1}$ is a norm in $\mathbf{H}_{\mathrm{per}}^1$ produced by the scalar product $(\cdot, \ \cdot)_{\mathbf{H}_{\mathrm{per}}^1}$.

3.4 Convergence Theorems

3.4.1 The Case of Rounded Edges

Consider a space $H_{\mathrm{per}}^1(\Omega_0) = \{ \varphi \in H^1(\Omega_0) : \ \varphi(x_1, \ 0, \ x_3) = \varphi(x_1, \ l, \ x_3), \ r < r_0 \}$ and a space of multivalued functions

$$\widetilde{\mathbf{H}}_{\mathrm{per}} = \{ \mathbf{p} = (p_0, \ p_1, \ p_2) \in \widetilde{\mathbf{H}} : \ p_0 \in H_{\mathrm{per}}^1(\Omega_0) \}$$

(see definition of $\widetilde{\mathbf{H}}$ in Sect. 2.3.1).

Theorem 3.3 *If $\gamma \in (-1, 0)$, then*

$$\left. \begin{array}{l} u_\varepsilon \xrightarrow{w} u_0 \quad \text{weakly in } H^1(\Omega_0), \\ \widetilde{u}_\varepsilon^{(1)} \xrightarrow{w} h_1 u_1 \ \text{weakly in } L^2(\Omega^{(1)}), \\ \widetilde{u}_\varepsilon^{(2)} \xrightarrow{w} h_2 u_2 \ \text{weakly in } L^2(\Omega^{(2)}) \end{array} \right\} \quad \text{as } \varepsilon \to 0, \tag{3.17}$$

where u_ε is the weak solution to problem (3.2) and the multivalued function $\mathbf{u} = (u_0, u_1, u_2) \in \widetilde{\mathbf{H}}_{\mathrm{per}}$ is a unique weak solution to the problem

$$\begin{cases} -\Delta_x u_0 = f_0, & x \in \Omega_0, \\ \partial^p_{x_2} u_0(x_1, 0, x_3) = \partial^p_{x_2} u_0(x_1, l, x_3), & p = 0, 1, \ x \in \partial\Omega_0 \cap \{r < r_0\}, \\ -\div_{\tilde x} (h_1(r)\nabla_{\tilde x} u_1) = h_1(r) f_0, & x \in \Omega^{(1)}, \\ \partial_\nu u_1 = 0, & x \in \partial\Omega^{(1)} \cap \{r = r_1\}, \\ -h_2\Delta_{\tilde x} u_2 + 2\kappa u_2 = h_2 f_0, & x \in \Omega^{(2)}, \\ u_2 = q_0, & x \in \partial\Omega^{(2)} \cap \{r = r_2\}, \\ u_0 = u_1 = u_2, & x \in \Omega', \\ \partial_r u_0 = h_1(r_0)\partial_r u_1 + h_2\partial_r u_2, & x \in \Omega'. \end{cases} \qquad (3.18)$$

To give the definition of a weak solution to problem (3.18), we introduce an anisotropic weighted Sobolev space of multivalued functions

$$\widetilde{\mathbf{H}}_{h_1,\mathrm{per}} = \Big\{ \mathbf{p} = (p_0,\, p_1,\, p_2) :\ p_0 \in H^1_{\mathrm{per}}(\Omega_0),\ p_1 \in \widetilde{H}_{h_1}(\Omega^{(1)}),\ p_2 \in \widetilde{H}^1(\Omega^{(2)}),$$

$$p_0|_{\Omega'} = p_1|_{\Omega'} = p_2|_{\Omega'} \Big\},$$

equipped with the scalar product

$$(\mathbf{p},\, \mathbf{q})_{\widetilde{\mathbf{H}}_{h_1,\mathrm{per}}} = \int_{\Omega_0} \nabla_x p_0 \cdot \nabla_x q_0 \, dx + \int_{\Omega^{(1)}} h_1(r)\nabla_{\tilde x} p_1 \cdot \nabla_{\tilde x} q_1 \, dx$$

$$+ \int_{\Omega^{(2)}} (h_2\nabla_{\tilde x} p_2 \cdot \nabla_{\tilde x} q_2 + 2\kappa p_2 q_2) \, dx,$$

and a linear functional

$$\mathsf{L}(\mathbf{p}) := \int_{\Omega_0} f_0 p_0 \, dx + \int_{\Omega^{(1)}} h_1(r) f_0 p_1 \, dx + h_2 \int_{\Omega^{(2)}} f_0 p_2 \, dx, \quad \mathbf{p} \in \widetilde{\mathbf{H}}_{h_1,\mathrm{per}}. \qquad (3.19)$$

According to (3.1), the embedding $\widetilde{\mathbf{H}}_{\mathrm{per}} \subset \widetilde{\mathbf{H}}_{h_1,\mathrm{per}}$ takes place.

Definition 3.3 [$\gamma \in (-1, 0)$] A multivalued function $\mathbf{u} = (u_0,\, u_1,\, u_2) \in \widetilde{\mathbf{H}}_{h_1,\mathrm{per}}$ is called a weak solution to problem (3.18) if $u_2|_{\partial\Omega^{(2)}\cap\{r=r_2\}} = q_0$ and

$$(\mathbf{u},\, \mathbf{p})_{\widetilde{\mathbf{H}}_{h_1,\mathrm{per}}} = \mathsf{L}(\mathbf{p}) \quad \forall \mathbf{p} = (p_0,\, p_1,\, p_2) \in \widetilde{\mathbf{H}}_{h_1,\mathrm{per}}, \quad p_2|_{\partial\Omega^{(2)}\cap\{r=r_2\}} = 0. \quad (3.20)$$

Proof **1.** Similarly as in the third part of the proof of Theorem 3.1, we prove that

$$\|u_\varepsilon\|_{H^2(\Omega^{(1)}_\varepsilon \cap R^{r_1}_{r_1-\delta_1})} \le c_0 \Big(\|f_\varepsilon\|_{L^2(\Omega^{(1)}_\varepsilon)} + \|u_\varepsilon\|_{H^1(\Omega^{(1)}_\varepsilon)} \Big).$$

According to this estimate, (3.1), Lemmas 3.2 and 3.4, the quantities

$$\|u_\varepsilon\|_{H^1(\Omega_0)}, \quad \|h_1^{-1}\widetilde{u}_\varepsilon^{(1)}\|_{L^2(\Omega^{(1)})}, \quad \|h_1^{-1}\widetilde{\partial_{x_k}u}_\varepsilon^{(1)}\|_{L^2(\Omega^{(1)})}, \quad \|\widetilde{u}_\varepsilon^{(i)}\|_{L^2(\Omega^{(i)})}, \quad \|\widetilde{\partial_{x_k}u}_\varepsilon^{(i)}\|_{L^2(\Omega^{(i)})}$$

are uniformly bounded with respect to ε ($i = 1, 2, k = 1, 2, 3$). Hence there exists a subsequence $\{\varepsilon'\} \subset \{\varepsilon\}$ (again denoted by $\{\varepsilon\}$) such that

$$\left.\begin{aligned}
u_\varepsilon &\xrightarrow{w} u_0 && \text{weakly in } H^1(\Omega_0), \\
h_1^{-1}\widetilde{u}_\varepsilon^{(1)} &\xrightarrow{w} u_1 && \text{weakly in } L^2(\Omega^{(1)}), \\
\widetilde{u}_\varepsilon^{(1)} &\xrightarrow{w} \widetilde{u}_1 && \text{weakly in } L^2(\Omega^{(1)}), \\
h_1^{-1}\widetilde{\partial_{x_k}u}_\varepsilon^{(1)} &\xrightarrow{w} u_{1,k} && \text{weakly in } L^2(\Omega^{(1)}), \\
\widetilde{\partial_{x_k}u}_\varepsilon^{(1)} &\xrightarrow{w} \widetilde{u}_{1,k} && \text{weakly in } L^2(\Omega^{(1)}), \\
\widetilde{u}_\varepsilon^{(2)} &\xrightarrow{w} \widetilde{u}_2 := h_2 u_2 && \text{weakly in } L^2(\Omega^{(2)}), \\
\widetilde{\partial_{x_k}u}_\varepsilon^{(2)} &\xrightarrow{w} \widetilde{u}_{2,k} := h_2 u_{2,k} && \text{weakly in } L^2(\Omega^{(2)}),
\end{aligned}\right\} \quad \text{as } \varepsilon \to 0 \quad (3.21)$$

where the limits remain unknown and will be defined later.

Obviously, $u_0 \in H^1_{\mathrm{per}}(\Omega_0)$. It is easy to verify that $\widetilde{u}_1 = h_1(r)u_1$ and $\widetilde{u}_{1,k} = h_1(r)u_{1,k}$ a.e. in $\Omega^{(1)}$, $k = 1, 2, 3$.

2. Repeating the assertions of the proof of Theorem 2.1, we can show that

- $u_{i,2} = 0$ a.e. in $\Omega^{(i)}$, $i = 1, 2$;
- there exist weak derivatives $\partial_{x_k}u_i = u_{i,k}$ for a.e. $x \in \Omega^{(i)}$, $i = 1, 2$, $k = 1, 3$;
- $u_0|_{\Omega'} = u_1|_{\Omega'} = u_2|_{\Omega'}$.

According to (2.28) and the Dirichlet boundary conditions on $\partial\Omega_\varepsilon^{(2)} \cap \{r = r_2\}$, the integral identity

$$\int_{\Omega_\varepsilon^{(2)}} r^{-1}\partial_r(u_\varepsilon\psi)\,dx = r_2^{-1}\int_{\partial\Omega_\varepsilon^{(2)}\cap\{r=r_2\}} u_\varepsilon\psi\,d\sigma_x = r_2^{-1}\int_{\partial\Omega_\varepsilon^{(2)}\cap\{r=r_2\}} q_\varepsilon\psi\,d\sigma_x$$

holds for any $\psi \in C^\infty(\overline{\Omega^{(2)}})$, $\psi|_{\Omega'} = 0$. Passing to the limit in this identity as $\varepsilon \to 0$ and taking into account (3.21), (3.4), and (2.16), we obtain

$$h_2\int_{\Omega^{(2)}} r^{-1}\partial_r(u_2^-\psi)\,dx = h_2 r_2^{-1}\int_{\partial\Omega^{(2)}\cap\{r=r_2\}} q_0\psi\,d\sigma_x \quad \forall\psi \in C^\infty(\overline{\Omega^{(2)}}), \ \psi|_{\Omega'} = 0,$$

whence integrating by parts, we derive that

$$h_2 r_2^{-1}\int_{\partial\Omega^{(2)}\cap\{r=r_2\}} u_2\psi\,d\sigma_x = h_2 r_2^{-1}\int_{\partial\Omega^{(2)}\cap\{r=r_2\}} q_0\psi\,d\sigma_x \quad \forall\psi \in C^\infty(\overline{\Omega^{(2)}}), \ \psi|_{\Omega'} = 0,$$

i.e.,

$$u_2|_{\partial\Omega^{(2)}\cap\{r=r_2\}} = q_0. \quad (3.22)$$

3. Using the functions u_0, u_1, u_2, we define a multivalued function $\mathbf{u} = (u_0, u_1, u_2)$. It is clear that $\mathbf{u} \in \overset{\circ}{\mathbf{H}}_{\mathrm{per}}$. With the help of (3.6) we rewrite (3.5) in a form

$$\int_{\Omega_0} \nabla_x u_\varepsilon \cdot \nabla_x p_0 \, dx + \sum_{i=1}^{2} \int_{\Omega^{(i)}} \widetilde{\nabla_x u_\varepsilon}^{(i)} \cdot \nabla_x p_i \, dx + \frac{2\kappa}{h_2} \int_{\Omega^{(2)}} \widetilde{u}_\varepsilon^{(2)} p_2 \, dx$$

$$- \frac{2\kappa\varepsilon}{h_2} \int_{\Omega_\varepsilon^{(2)}} Y_1\left(\frac{x_2}{\varepsilon}\right) \partial_{x_2}(u_\varepsilon p_2) \, dx = \int_{\Omega_0} f_\varepsilon p_0 \, dx + \sum_{i=1}^{2} \int_{\Omega^{(i)}} \chi_{\Omega_\varepsilon^{(i)}} f_\varepsilon p_i \, dx,$$

$$(3.23)$$

where $\mathbf{p} = (p_0, \, p_1, \, p_2)$ is an arbitrary function from

$$\mathbf{C}_{\text{per}}^\infty(\partial\Omega^{(2)} \cap \{r = r_2\}) := \{\mathbf{p} = (p_0, \, p_1, \, p_2) \in \widetilde{\mathbf{H}}_{\text{per}}^1 : p_0 \in C^\infty(\overline{\Omega}_0), \ p_1 \in C^\infty(\overline{\Omega^{(1)}}),$$

$$p_2 \in C^\infty(\overline{\Omega^{(2)}}, \partial\Omega^{(2)} \cap \{r = r_2\})\}.$$

Passing to the limit in (3.23) and taking (2.3), (3.21), and (2.16) into account, we obtain the identity

$$(\mathbf{u}, \, \mathbf{p})_{\widetilde{\mathbf{H}}_{h_1,\text{per}}} = \mathsf{L}(\mathbf{p}) \quad \forall \, \mathbf{p} \in \mathbf{C}_{\text{per}}^\infty(\partial\Omega^{(2)} \cap \{r = r_2\}),$$

where the functional L is defined in (3.19). Since $\mathbf{C}_{\text{per}}^\infty(\partial\Omega^{(2)} \cap \{r = r_2\})$ is dense in the space of functions from $\widetilde{\mathbf{H}}_{h_1,\text{per}}$ such that $p_2|_{\partial\Omega^{(2)} \cap \{r=r_2\}} = 0$, we arrive at (3.20).

4. Let us verify that the functional L is bounded, i.e.,

$$|\mathsf{L}(\mathbf{p})| \le c_5 \|\mathbf{p}\|_{\widetilde{\mathbf{H}}_{h_1,\text{per}}} \quad \forall \, \mathbf{p} = (p_0, \, p_1, \, p_2) \in \widetilde{\mathbf{H}}_{h_1,\text{per}}, \quad p_2|_{\partial\Omega^{(2)} \cap \{r=r_2\}} = 0,$$

where $\|\cdot\|_{\widetilde{\mathbf{H}}_{h_1,\text{per}}}$ is a norm produced by the scalar product $(\cdot, \, \cdot)_{\widetilde{\mathbf{H}}_{h_1,\text{per}}}$. This together with (3.22) will imply that $\mathbf{u}$ is a unique weak solution to problem (3.18).

Obvious inequalities

$$\int_{\Omega_0} p_0^2 \, dx \le c_1 \left(\|p_0\|_{L^2(\Omega')}^2 + \int_{\Omega_0} |\nabla_{\tilde{x}} p_0|^2 \, dx \right)$$

and

$$\|p_0\|_{L^2(\Omega')}^2 = \|p_1\|_{L^2(\Omega')}^2 = \|p_2\|_{L^2(\Omega')}^2 \le c_2 \int_{\Omega^{(2)}} (|\nabla_{\tilde{x}} p_2|^2 + p_2^2) \, dx$$

provide that

$$\left| \int_{\Omega_0} f_0 p_0 \, dx \right| \le c_3 \|f_0\|_{L^2(\Omega_0)} \|\mathbf{p}\|_{\widetilde{\mathbf{H}}_{h_1,\text{per}}}. \tag{3.24}$$

The second summand in (3.19) is estimated with the help of the Cauchy–Schwartz–Bunyakovsky inequality:

$$\left| \int_{\Omega^{(1)}} h_1(r) f_0 p_1 \, dx \right| \le c_4 \| f_0 \|_{L^2(r_1)} \left(\int_{\Omega^{(1)}} h_1(r) p_1^2 \, dx \right)^{\frac{1}{2}}.$$

It is easy to see that

$$\int_{\Omega^{(1)}} h_1(r) p_1^2 \, dx \le c_5 \left(\| p_1 \|_{L^2(\Omega')}^2 + \int_{\Omega^{(1)}} h_1(r) \int_{r_0}^{r} (\partial_\xi p_1 |_{r=\xi})^2 \, d\xi \, dx \right).$$

Using (3.1), we derive the estimate

$$\int_{\Omega^{(1)}} \int_{r}^{r_1} h_1(\xi) (\partial_r p_1)^2 \, d\xi \, dx \le c_5 \int_{\Omega^{(1)}} h_1(r) |\nabla_{\tilde{x}} p_1|^2 \, dx.$$

The last inequality and the Fubini theorem imply that

$$\int_{\Omega^{(1)}} h_1(r) \int_{r_0}^{r} (\partial_\xi p_1 |_{r=\xi})^2 \, d\xi \, dx \le c_6 \int_{\Omega^{(1)}} h_1(r) |\nabla_{\tilde{x}} p_1|^2 \, dx.$$

Taking into account (3.24) and the obtained inequalities, we conclude that

$$\left| \int_{\Omega^{(1)}} h_1(r) f_0 p_1 \, dx \right| \le c_7 \| f_0 \|_{L^2(\Omega^{(1)})} \| \mathbf{p} \|_{\tilde{\mathbf{H}}_{h_1, \text{per}}}. \tag{3.25}$$

Understandably that the third summand in identity (3.19) can be estimated by $c_8 \| f_0 \|_{L^2(\Omega^{(2)})} \| \mathbf{p} \|_{\tilde{\mathbf{H}}_{h_1, \text{per}}}$. This estimate together with (3.24) and (3.25) give the boundedness of L.

5. Since all of the above assertions remain valid for an arbitrary subsequence $\{\varepsilon'\}$ chosen at the beginning of the proof, the uniqueness of a weak solution to problem (3.18) implies that relations (3.17) hold for the whole sequence $\{\varepsilon\}$.

3.4.2 The Case of Wedge Edges

Theorem 3.4 If $\gamma = 0$, then the extension $\mathbf{E}_{u_\varepsilon} = \left(u_\varepsilon|_{\Omega_0}, E_{u_\varepsilon}^{(1)}|_{\Omega^{(1)}}, E_{u_\varepsilon}^{(2)}|_{\Omega^{(2)}} \right)$ defined in (3.15) of the weak solution u_ε to problem (3.2) satisfies the relations

$$\mathbf{E}_{u_\varepsilon} \xrightarrow{w} \mathbf{u} \quad \text{weakly in } \mathbf{H}_{\text{per}}^1 \text{ as } \varepsilon \to 0, \tag{3.26}$$

$$\lim_{\mu \to 0} \lim_{\varepsilon \to 0} \int_{\partial \Omega^{(1)} \cap \{r=r_1\}} (E_{u_\varepsilon}^{(1)}|_{r=r_1-\mu} - u_1|_{r=r_1})^2 \, d\sigma_x = 0, \tag{3.27}$$

where the multivalued function $\mathbf{u} = (u_0, u_1, u_2) \in \mathbf{H}_{\text{per}}^1$ is a unique weak solution to the problem

$$\begin{cases} -\Delta_x u_0 = f_0, & x \in \Omega_0, \\ \partial_{x_2}^p u_0(x_1, 0, x_3) = \partial_{x_2}^p u_0(x_1, l, x_3), & p = 0, 1, \ x \in \partial\Omega_0 \cap \{r < r_0\}, \\ - \div_{\bar{x}}\, (h_1(r)\nabla_{\bar{x}} u_1) = h_1(r) f_0, & x \in \Omega^{(1)}, \\ -h_2 \Delta_{\bar{x}} u_2 + 2\kappa u_2 = h_2 f_0, & x \in \Omega^{(2)}, \\ u_2 = q_0, & x \in \partial\Omega^{(2)} \cap \{r = r_2\}, \\ u_0 = u_1 = u_2, & x \in \Omega', \\ \partial_r u_0 = h_1(r_0)\partial_r u_1 + h_2 \partial_r u_2, & x \in \Omega'. \end{cases} \tag{3.28}$$

Remark 3.4 The convergence (3.27) was introduced by Mikhailov in [116, Ch. IV].

Remark 3.5 If $\gamma \geq 0$, then due to Lemma 3.5, the embedding

$$C^\infty\big(\overline{\Omega^{(1)}}\big) \subset \widetilde{H}_{h_1}(\Omega^{(1)}, \partial\Omega^{(1)} \cap \{r = r_1\}) \tag{3.29}$$

takes place. This means that $\widetilde{H}_{h_1}(\Omega^{(1)}) = \widetilde{H}_{h_1}(\Omega^{(1)}, \partial\Omega^{(1)} \cap \{r = r_1\})$. Thus, the definition of a weak solution to problem (3.28) coincides with Definition 3.3. As a consequence, problem (3.28) has a unique weak solution.

The embedding (3.29) also implies that there is no need for any boundary conditions on $\partial\Omega^{(1)} \cap \{r = r_1\}$ for uniqueness of a weak solution to problem (3.28) (the similar situation was in [133, §3] and [123]).

However, in the case $\gamma = 0$ we will prove in Theorem 3.4 that the weak solution belongs to $\mathbf{H}^1_{\mathrm{per}}$ and, as a consequence, has a finite trace on $\partial\Omega^{(1)} \cap \{r = r_1\}$ (see (3.27)).

Proof **1.** With the help of (3.16), we can prove that there exists a subsequence $\{\varepsilon'\} \subset \{\varepsilon\}$ (again denoted by $\{\varepsilon\}$) such that

$$\left. \begin{array}{ll} \mathbf{E}_{u_\varepsilon} \xrightarrow{\ w\ } \mathbf{u} := (u_0, u_1, u_2) & \text{weakly in } \mathbf{H}^1_{\mathrm{per}}, \\ \chi_{\Omega_\varepsilon^{(i)}} \partial_{x_k} E_{u_\varepsilon}^{(i)} \xrightarrow{\ w\ } u_{i,k} & \text{weakly in } L^2(\Omega^{(i)}) \end{array} \right\} \quad \text{as } \varepsilon \to 0, \tag{3.30}$$

where u_0, u_i, $u_{i,k}$, $k = 1, 2, 3$, $i = 1, 2$, will be defined later. From (3.4) and Theorem 3.2, it follows that $u_2|_{\partial\Omega^{(2)}\cap\{r=r_2\}} = q_0$.

Compactness of the embedding $H^1(\Omega^{(i)}) \subset L^2(\Omega^{(i)})$ and relations (2.16) and (3.30) imply that

$$\left. \begin{array}{ll} \chi_{\Omega_\varepsilon^{(1)}} E_{u_\varepsilon}^{(1)} \xrightarrow{\ w\ } h_1 u_1 & \text{weakly in } L^2(\Omega^{(1)}), \\ \chi_{\Omega_\varepsilon^{(2)}} E_{u_\varepsilon}^{(2)} \xrightarrow{\ w\ } h_2 u_2 & \text{weakly in } L^2(\Omega^{(2)}) \end{array} \right\} \quad \text{as } \varepsilon \to 0. \tag{3.31}$$

2. Similarly as in the second part of the proof of Theorem 2.1, we show that $u_{i,2} = 0$ a.e. in $\Omega^{(i)}$, $i = 1, 2$. Next let us find $u_{i,1}$ and $u_{i,3}$, $i = 1, 2$.

By the same way as the identity (2.27) was proved, we deduce

$$\int_{\Omega^{(1)}} \chi_{\Omega^{(1)}_\varepsilon} \partial_{x_k} E^{(1)}_{u_\varepsilon} \psi \, dx = - \int_{\Omega^{(1)}} \chi_{\Omega^{(1)}_\varepsilon} E^{(1)}_{u_\varepsilon} (\partial_{x_k} \psi + \partial_{x_k} \ln h_1(r) \psi) \, dx$$

$$+ \varepsilon \int_{\Omega^{(1)}_\varepsilon} Y_1 \left(\frac{x_2}{\varepsilon} \right) \partial_{x_k} \ln h_1(r) \partial_{x_2}(u_\varepsilon \psi) \, dx \quad \forall \psi \in C_0^\infty(\Omega^{(1)}), \ k = 1, \, 3.$$

Passing to the limit in this identity as $\varepsilon \to 0$ and taking (3.30) and (3.31) into account, we get the identity

$$\int_{\Omega^{(1)}} u_{1,k} \psi \, dx = - \int_{\Omega^{(1)}} u_1 \partial_{x_k}(h_1(r) \psi) \, dx \quad \forall \psi \in C_0^\infty(\Omega^{(1)}), \ k = 1, \, 3,$$

which implies that $u_{1,k} = h_1(r) \partial_{x_k} u_1$ a.e. in $\Omega^{(1)}$, $k = 1, \, 3$. Similarly with the help of (3.6) we have that $u_{2,k} = h_2 \partial_{x_k} u_2$ a.e. in $\Omega^{(2)}$, $k = 1, \, 3$.

3. Let us prove convergence (3.27). Let μ be any fixed number from $(0, \frac{r_1 - r_0}{4})$. Properties of the trace operator and the first relation in (3.30) imply that

$$E^{(1)}_{u_\varepsilon}|_{\{r = r_1 - \mu\}} \longrightarrow u_1|_{\{r = r_1 - \mu\}} \quad \text{strongly in } L^2(\{r = r_1 - \mu\}) \text{ as } \varepsilon \to 0.$$

This convergence and the inequality

$$(u_1|_{r=r_1-\mu} - u_1|_{r=r_1})^2 \le \mu \int_{r_1-\mu}^{r_1} (\partial_r u_1)^2 \, dr$$

provide convergence (3.27):

$$\|u_1|_{r=r_1-\mu} - u_1|_{r=r_1}\|_{L^2(\partial\Omega^{(1)} \cap \{r=r_1\})} \le c_0 \mu^{\frac{1}{2}} \|u_1\|_{H^1(\Omega^{(1)})} \longrightarrow 0 \quad \text{as } \mu \to 0.$$

4. Utilizing (3.6) and the extension $\mathbf{E}_{u_\varepsilon}$, we rewrite identity (3.5) as follows

$$\int_{\Omega_0} \nabla_x u_\varepsilon \cdot \nabla_x p_0 \, dx + \sum_{i=1}^{2} \int_{\Omega^{(i)}} \chi_{\Omega^{(i)}_\varepsilon} \nabla_x E^{(i)}_{u_\varepsilon} \cdot \nabla_x p_i \, dx + \frac{2\kappa}{h_2} \int_{\Omega^{(2)}} \chi_{\Omega^{(2)}_\varepsilon} E^{(2)}_{u_\varepsilon} p_2 \, dx$$

$$- \frac{2\kappa\varepsilon}{h_2} \int_{\Omega^{(2)}_\varepsilon} Y_1 \left(\frac{x_2}{\varepsilon} \right) \partial_{x_2}(u_\varepsilon p_2) \, dx = \int_{\Omega_0} f_\varepsilon p_0 \, dx + \sum_{i=1}^{2} \int_{\Omega^{(i)}} \chi_{\Omega^{(i)}_\varepsilon} f_\varepsilon p_i \, dx$$

$$\tag{3.32}$$

with arbitrary multivalued test function $\mathbf{p} = (p_0, \ p_1, \ p_2)$ from the space

$$\mathbf{C}^\infty_{\mathrm{per}}(\partial\Omega^{(1)} \cap \{r = r_1\}, \ \partial\Omega^{(2)} \cap \{r = r_2\}) := \{\mathbf{p} = (p_0, \ p_1, \ p_2) \in \mathbf{H}^1_{\mathrm{per}} :$$

$$p_0 \in C^\infty(\overline{\Omega_0}), \ p_i \in C^\infty(\overline{\Omega^{(i)}}, \partial\Omega^{(i)} \cap \{r = r_i\}), \ i = 1, \, 2\}.$$

Taking (2.3), (2.16), (3.30), and (3.31) into account and passing to the limit in (3.32), we derive the following identity:

$$(\mathbf{u},\ \mathbf{p})_{\widetilde{\mathbf{H}}_{h_1,\mathrm{per}}} = \mathrm{L}(\mathbf{p}) \quad \forall\,\mathbf{p} \in \mathbf{C}^{\infty}_{\mathrm{per}}\big(\partial\Omega^{(1)} \cap \{r = r_1\},\ \partial\Omega^{(2)} \cap \{r = r_2\}\big).$$

Since $\mathbf{u} \in \mathbf{H}^1_{\mathrm{per}}$, then the last identity remains valid for all $\mathbf{p} = (p_0,\ p_1,\ p_2) \in \widetilde{\mathbf{H}}_{h_1,\mathrm{per}}$ such that $p_2|_{\partial\Omega^{(2)}\cap\{r=r_2\}} = 0$ (see Remark 3.5). Thus, $\mathbf{u} \in \mathbf{H}^1_{\mathrm{per}}$ is a unique weak solution to problem (3.28).

5. Since all of the above assertions remain valid for arbitrary subsequence $\{\varepsilon'\}$ chosen at the beginning of the proof, the uniqueness of a weak solution to problem (3.28) implies that relations (3.26) and (3.27) hold for the whole sequence $\{\varepsilon\}$.

3.4.3   The Case of Very Sharp Edges

The existence and uniqueness of the weak solution to problem (3.2) in the case $\gamma > 0$ and some of its properties are justified in Remark 3.5.

We introduce a space $L^2_\alpha(\Omega^{(1)}) := \{\varphi \in L^2(\Omega^{(1)}) : \ \|(r_1 - r)^{\frac{\alpha}{2}}\varphi\|_{L^2(\Omega^{(1)})} < \infty\}$, where $\alpha \in \mathbb{R}$, and a weighted Sobolev space of multivalued functions

$$\begin{aligned}
\widetilde{\mathbf{H}}_\gamma := \big\{ \mathbf{p} = (p_0,\ p_1,\ p_2) : \ & p_0 \in H^1_{\mathrm{per}}(\Omega_0),\\
& p_1 \in L^2_{2\gamma}(\Omega^{(1)}),\ \exists\,\partial_{x_k}p_1 \in L^2_{2(1+\gamma)}(\Omega^{(1)}),\ k = 1,\,3,\\
& p_2 \in \widetilde{H}^1(\Omega^{(2)}),\ p_0|_{\Omega'} = p_1|_{\Omega'} = p_2|_{\Omega'} \big\}.
\end{aligned}$$

Theorem 3.5 *If $\gamma > 0$, then there exists a subsequence $\{\varepsilon'\} \subset \{\varepsilon\}$ such that the weak solution $u_{\varepsilon'}$ to problem (3.2) satisfies the relations*

$$\left.\begin{aligned}
u_{\varepsilon'} &\xrightarrow{\ w\ } u_0 && \text{weakly in } H^1(\Omega_0),\\
\widetilde{u}^{(1)}_{\varepsilon'} &\xrightarrow{\ w\ } h_1 u_1 && \text{weakly in } L^2(\Omega^{(1)}),\\
\widetilde{u}^{(2)}_{\varepsilon'} &\xrightarrow{\ w\ } h_2 u_2 && \text{weakly in } L^2(\Omega^{(2)})
\end{aligned}\right\} \quad \text{as } \varepsilon' \to 0, \tag{3.33}$$

where the multivalued function $\mathbf{u} := (u_0,\ u_1,\ u_2)$ belongs to the space $\widetilde{\mathbf{H}}_\gamma$ and satisfies the following relations: $u_2|_{\partial\Omega^{(2)}\cap\{r=r_2\}} = q_0$ and

$$(\mathbf{u},\ \mathbf{p})_{\widetilde{\mathbf{H}}_{h_1,\mathrm{per}}} = \mathrm{L}(\mathbf{p}) \tag{3.34}$$

for all $\mathbf{p} = (p_0,\ p_1,\ p_2) \in \widetilde{\mathbf{H}}_{h_1,\mathrm{per}}$ such that $p_1 = 0$ in some neighborhood of $\partial\Omega^{(1)} \cap \{r = r_1\}$ (not necessarily the same for all functions) and $p_2|_{\partial\Omega^{(2)}\cap\{r=r_2\}} = 0$.

Proof Similarly as in [123, Lemma 4.1] we can prove that

$$\|(r_1 - r)^{-1}\varphi\|_{L^2(\Omega_\varepsilon^{(1)})} \le c_0 \|\varphi\|_{H^1(\Omega_\varepsilon)} \quad \forall\,\varphi \in H^1_{\mathrm{per}}(\Omega_\varepsilon,\ \partial\Omega^{(1)}_\varepsilon \cap \{r = r_1\}).$$

According to Lemma 3.4 and the last estimate the quantities

$$\|u_\varepsilon\|_{H^1(\Omega_0)}, \quad \|(r_1 - r)^{-1}\widetilde{u}_\varepsilon^{(1)}\|_{L^2(\Omega^{(1)})}, \quad \|\widetilde{u}_\varepsilon^{(i)}\|_{L^2(\Omega^{(i)})}, \quad \|\widetilde{\partial_{x_k} u}_\varepsilon^{(i)}\|_{L^2(\Omega^{(i)})},$$

where $i = \{1, 2\}$, $k \in \{1, 2, 3\}$, are bounded uniformly with respect to ε. Thus, there exists a subsequence $\{\varepsilon'\} \subset \{\varepsilon\}$ (again denoted by $\{\varepsilon\}$) such that

$$\left.\begin{aligned}
u_\varepsilon &\xrightarrow{w} u_0 && \text{weakly in } H^1(\Omega_0), \\
(r_1 - r)^{-1}\widetilde{u}_\varepsilon^{(1)} &\xrightarrow{w} w_1 && \text{weakly in } L^2(\Omega^{(1)}), \\
\widetilde{u}_\varepsilon^{(i)} &\xrightarrow{w} \widetilde{u}_i := h_i u_i && \text{weakly in } L^2(\Omega^{(i)}), \\
\widetilde{\partial_{x_k} u}_\varepsilon^{(i)} &\xrightarrow{w} \widetilde{u}_{i,k} := h_i u_{i,k} && \text{weakly in } L^2(\Omega^{(i)}),
\end{aligned}\right\} \quad \text{as } \varepsilon \to 0. \quad (3.35)$$

Here the limits remain unknown and will be defined later.

By the same assertions as in the proof of Theorem 2.1 we can show that

- $u_{i,2} = 0$ a.e. in $\Omega^{(i)}$, there exist weak derivatives $\partial_{x_k} u_i$ and $u_{i,k} = \partial_{x_k} u_i$ a.e. in $\Omega^{(i)}$, $k = 1, 3$, $i = 1, 2$;
- $u_0|_{\Omega'} = u_1|_{\Omega'} = u_2|_{\Omega'}$;
- $u_0(x_1, 0, x_3) = u_0(x_1, l, x_3)$, $x \in \partial\Omega_0 \cap \{r < r_0\}$;
- $u_2|_{\partial\Omega^{(2)} \cap \{r=r_2\}} = q_0$.

From (3.35) and (3.1) it follows that $w_1 = (r_1 - r)^{-1}h_1(r)u_1 \in L^2(\Omega^{(1)})$, i.e., $u_1 \in L^2_{2\gamma}(\Omega^{(1)})$. Thus, $\partial_{x_k} u_1 \in L^2_{2(1+\gamma)}(\Omega^{(1)})$, $k = 1, 3$. Thus, the multivalued function $\mathbf{u} = (u_0, u_1, u_2)$ belongs to the weighted space $\widetilde{\mathbf{H}}_\gamma$.

Consider arbitrary function $\mathbf{p} = (p_0, p_1, p_2) \in \widetilde{\mathbf{H}}_{h_1,\text{per}}$ such that $p_1 = 0$ in some neighborhood of $\partial\Omega^{(1)} \cap \{r = r_1\}$ (not necessarily the same for all functions) and $p_2|_{\partial\Omega^{(2)} \cap \{r=r_2\}} = 0$. Then we define a test function

$$\Phi(x) = \begin{cases} p_0(x), & x \in \Omega_0, \\ p_1(x), & x \in \Omega_\varepsilon^{(1)}, \\ p_2(x), & x \in \Omega_\varepsilon^{(2)}. \end{cases}$$

Clearly, $\Phi \in H^1_{\text{per}}(\Omega_\varepsilon, \partial\Omega_\varepsilon^{(1)} \cap \{r = r_1\})$ and $\Phi|_{\partial\Omega_\varepsilon^{(2)} \cap \{r=r_2\}} = 0$.

We rewrite the identity (3.5) with the test function Φ and obtain (3.23). Sending ε to zero in (3.23), we obtain identity (3.34) for the multivalued function $\mathbf{u} \in \widetilde{\mathbf{H}}_\gamma$.

Remark 3.6 If $\mathbf{u} \in \widetilde{\mathbf{H}}_{h_1,\text{per}}$, then $\mathbf{u}$ is the unique weak solution to problem (3.28)) (see [133, §3]), and consequently convergence (3.33) hold for the whole sequence $\{\varepsilon\}$. Unfortunately, we could not prove that $\mathbf{u} \in \widetilde{\mathbf{H}}_{h_1,\text{per}}$.

3.5 Conclusions to this Chapter

The homogenized problems (3.18) and (3.28) are boundary-value problems for degenerate elliptic equations since the coefficient h_1 vanishes on the boundary

$\partial \Omega^{(1)} \cap \{r = r_1\}$. The parameter γ that described the degeneracy order plays a crucial role in the asymptotic behavior of the solution u_ε to problem (3.2).

1. In the case $\gamma \in (-1, 0)$, i.e., the first-level thin disks have *rounded edges* (cf. Fig. 3.2 a), the Neumann conditions on these edges are transformed into the same Neumann condition on $\partial \Omega^{(1)} \cap \{r = r_1\}$. In fact, the homogenized problem (3.18) coincides with a homogenized problem as if $h_1(r) > 0$ for all $\in [r_0, r_1]$ (cf., for example, the homogenized problem in [31]). This is confirmed also by the fact that the component u_1 of the multivalued solution to problem (3.18) belongs to the weighted Sobolev space $\widetilde{H}_{h_1}(\Omega^{(1)})$ and as was shown in [133] each function from this space has the finite trace on $\partial \Omega^{(1)} \cap \{r = r_1\}$ if $\gamma \in (-1, 0)$.

2. If $\gamma \geq 0$, no boundary condition on $\partial \Omega^{(1)} \cap \{r = r_1\}$ is required for establishing the existence and uniqueness of a solution to the homogenized problem (3.28) (cf. [133]). Moreover, some functions from the corresponding weighted Sobolev space may not have finite trace on $\partial \Omega^{(1)} \cap \{r = r_1\}$ (see Lemma 3.5). However there is a significant difference between two cases $\gamma = 0$ and $\gamma > 0$.

   If $\gamma = 0$, i.e., the first-level thin disks have *wedge edges* (cf. Fig. 3.2 b), then the solution **u** to the homogenized problem (3.28) belongs to the "regular" space $\mathbf{H}^1_{\mathrm{per}}$ and, as a consequence, the component u_1 has finite trace on $\partial \Omega^{(1)} \cap \{r = r_1\}$ (see Theorem 3.4). In addition, the convergence of the traces (3.27) in sense of Mikhailov is established on this part of the boundary.

3. In the case of *very sharp edges* ($\gamma > 0$) we cannot assert that the component u_1 has a trace on $\partial \Omega^{(1)} \cap \{r = r_1\}$ (cf. Remark 3.3) and its trace can even tend to infinity as $r \to r_1$ (see Lemma 3.5). As a consequence, we cannot establish uniqueness of a solution of the identity (3.34) in the space $\widetilde{\mathbf{H}}_\gamma$ and convergence (3.33) for the whole sequence (see Remark 3.6).

4. From physical point of view, the absence of conditions on $\partial \Omega^{(1)} \cap \{r = r_1\}$ if $\gamma \geq 0$ means that the heat dissipates into environment while reaching the spikes. This is not true in the case $\gamma \in (-1, 0)$. Thus, our results mathematically justify the following physical effect: heat radiators, shaped like a thick junction, radiate more heat if their thin joint domains have sharp edges.

5. The homogeneous Robin boundary conditions on the surfaces of the thin discs from the second level are transformed into the new summand $2\kappa u_2$ in the corresponding partial differential equation (similarly as in Chap. 2). Also, we studied the behavior of the inhomogeneous Dirichlet conditions on edges as $\varepsilon \to 0$.

6. In this chapter, we have focused on studying the influence of the geometric structure of thin disks from the first level on the asymptotic behavior of the solution. Therefore, to avoid additional cumbersome calculations, we consider the homogeneous Neumann boundary condition. From calculations made here and in Chap. 2, it follows that the Robin boundary condition $\partial_\nu u_\varepsilon + \varepsilon k_1 u_\varepsilon = 0$ can be also considered (a term $2k_1 u_1$ appears in the corresponding differential equation).

Chapter 4
Homogenization of Semilinear Parabolic Problems

Two semilinear parabolic problems in a thick two-level junction are considered. The first one is with different alternating nonlinear Robin boundary conditions and the other is with alternating Robin and Dirichlet conditions. In both problems, the passage to the limit is accompanied by a special intensity factor ε^{α} in the nonlinear term of the Robin conditions. The case of a big boundary interaction ($\alpha < 1$), which was not studied for the linear elliptic problem in Chap. 2, is examined. We establish qualitatively different cases in the asymptotic behavior of the solutions to those problems as $\varepsilon \to 0$ depending on the value of α. The convergence theorems are proved using the method of special integral identity and zero-extension operators. The limits of nonlinear terms are found with the help of the Minty–Browder method.

4.1 Problem with Alternating Robin Boundary Conditions

4.1.1 Statement of the Problem

Let T be a fixed positive number. We consider the following boundary-value problem in the thick junction Ω_{ε} described in Sect. 2.1:

$$\begin{cases} \partial_t u_{\varepsilon}(t, x) - \Delta_x u_{\varepsilon}(t, x) + k_0(u_{\varepsilon}(t, x)) = f_{\varepsilon}, & (t, x) \in (0, T) \times \Omega_{\varepsilon}, \\ \partial_{\nu} u_{\varepsilon}(t, x) + \varepsilon^{\alpha} \kappa_1(u_{\varepsilon}(t, x)) = \varepsilon^{\beta} g_{\varepsilon}(t, x), & (t, x) \in (0, T) \times \partial\Omega_{\varepsilon}^{(1)} \cap \{r > r_0\}, \\ \partial_{\nu} u_{\varepsilon}(t, x) + \varepsilon \kappa_2(u_{\varepsilon}(t, x)) = \varepsilon^{\beta} g_{\varepsilon}(t, x), & (t, x) \in (0, T) \times \partial\Omega_{\varepsilon}^{(2)} \cap \{r > r_0\}, \\ \partial_{\nu} u_{\varepsilon}(t, x) = 0, & (t, x) \in (0, T) \times \partial\Omega_{\varepsilon} \cap \{r = r_0\}, \\ \partial_{\nu} u_{\varepsilon}(t, x) = q_{\varepsilon}(t, x), & (t, x) \in (0, T) \times \partial\Omega_0 \cap \{r < r_0\}, \\ u_{\varepsilon}(0, x) = 0, & x \in \Omega_{\varepsilon}. \end{cases} \quad (4.1)$$

© The Author(s), under exclusive license to Springer Nature Switzerland AG 2019
T. Mel'nyk and D. Sadovyi, *Multiple-Scale Analysis of Boundary-Value Problems in Thick Multi-Level Junctions of Type 3:2:2*, SpringerBriefs in Mathematics, https://doi.org/10.1007/978-3-030-35537-1_4

Here $\alpha \in \mathbb{R}$, $\beta \geq 1$ are perturbation parameters. The given functions f_ε, g_ε, k_0, κ_1, κ_2, and q_ε satisfy the following conditions:

- the functions f_ε, $f_0 \in L^2((0, T) \times \Omega_0 \cup \Omega^{(2)})$, and

$$f_\varepsilon \longrightarrow f_0 \quad \text{strongly in } L^2((0, T) \times \Omega_0 \cup \Omega^{(2)}) \text{ as } \varepsilon \to 0; \qquad (4.2)$$

- the functions g_ε, $g_0 \in L^2(0, T; H^1(\Omega^{(2)}))$, and

$$g_\varepsilon \xrightarrow{w} g_0 \quad \text{weakly in } L^2(0, T; H^1(\Omega^{(2)})) \text{ as } \varepsilon \to 0; \qquad (4.3)$$

- the functions q_ε, $q_0 \in L^2((0, T) \times \partial\Omega_0 \cap \{r < r_0\})$, and

$$q_\varepsilon \xrightarrow{w} q_0 \quad \text{weakly in } L^2((0, T) \times \partial\Omega_0 \cap \{r < r_0\}) \text{ as } \varepsilon \to 0; \qquad (4.4)$$

- the functions $k_0, \kappa_m : \mathbb{R} \mapsto \mathbb{R}$ are Lipschitz-continuous, i.e., $k_0, \kappa_m \in W^{1,\infty}_{\text{loc}}(\mathbb{R})$, and there exist positive constants C_0 and C_1 such that

$$C_0 \leq k_0(s) \leq C_1, \qquad C_0 \leq \kappa'_m(s) \leq C_1 \ (m = 1, 2) \quad \text{for a.e. } s \in \mathbb{R}; \qquad (4.5)$$

- if $\alpha < 1$, then we additionally suppose that

$$\kappa_1(0) = 0. \qquad (4.6)$$

Remark 4.1 Robin's boundary condition means some interaction of a physical process occurring inside of a material with the external environment through its surface. In fact, very small activity holds always on the surface of some material (therefore Robin boundary conditions, in particular nonlinear ones, are more natural for applied mathematical problems). Many physical processes, especially in chemistry and medicine, have a monotonous nature. Therefore, it is natural to impose special monotonous assumptions for nonlinear terms. In our case, we propose simple assumptions (4.5) that are easy to verify. For instance, the functions

$$k(s) = \lambda s + \sin s \ (\lambda > 1); \quad k(s) = s + \arctan s; \quad k(s) = \frac{\lambda s}{1 + \nu s} \ (\lambda, \nu > 0)$$

satisfy (4.5). The last one corresponds to the Michaelis–Menten hypothesis in biochemical reactions and to the Langmuir kinetics adsorption models (e.g. [125]).

The case $\alpha < 1$ corresponds to relatively high (of order ε^α) heat transduction on the surfaces of the thin discs from the first level. The condition $\kappa_1(0) = 0$ means that if the temperature is zero on the boundary at some moment, then there is no transduction on the surfaces. In many applications, this condition is called *zero-absorption condition*.

Remark 4.2 Doubtless both the functions k_0, κ_1, κ_2 may also depend on x and t (cf. [92]). However, we have omitted this dependence to avoid cumbersome formulas.

Consider a space

$$W_T(\Omega_\varepsilon) = \left\{ \varphi \in L^2(0, T; H^1(\Omega_\varepsilon)) : \partial_t\varphi := \varphi' \in L^2(0, T; (H^1(\Omega_\varepsilon))^*) \right\}.$$

It is known (e.g., [42, Theorem 1.17]) that $W_T(\Omega_\varepsilon) \subset C([0, T]; L^2(\Omega_\varepsilon))$.

Definition 4.1 A function $u_\varepsilon \in W_T(\Omega_\varepsilon)$ is called a weak solution to problem (4.1) if $u_\varepsilon(0, x) = 0$ and the following integral identity takes place:

$$\mathsf{B}_{1,\varepsilon}(u_\varepsilon, \varphi) = \mathsf{L}_{1,\varepsilon}(\varphi) \quad \forall \varphi \in W_T(\Omega_\varepsilon), \tag{4.7}$$

where

$$\mathsf{B}_{1,\varepsilon}(u_\varepsilon, \varphi)$$
$$:= \int_{\Omega_\varepsilon} (u_\varepsilon\varphi)|_{t=T} \, dx + \int_0^T \left(-\langle \partial_t\varphi, u_\varepsilon \rangle_{H^1(\Omega_\varepsilon)} + \int_{\Omega_\varepsilon} (\nabla_x u_\varepsilon \cdot \nabla_x\varphi + k_0(u_\varepsilon)\varphi) \, dx \right.$$
$$\left. + \varepsilon^\alpha \int_{\partial\Omega_\varepsilon^{(1)} \cap \{r > r_0\}} \kappa_1(u_\varepsilon)\varphi \, d\sigma_x + \varepsilon \int_{\partial\Omega_\varepsilon^{(2)} \cap \{r > r_0\}} \kappa_2(u_\varepsilon)\varphi \, d\sigma_x \right) dt,$$

$$\mathsf{L}_{1,\varepsilon}(\varphi) := \int_0^T \left(\int_{\Omega_\varepsilon} f_\varepsilon\varphi \, dx + \int_{\partial\Omega_0 \cap \{r < r_0\}} q_\varepsilon\varphi \, d\tilde{x} + \varepsilon^\beta \int_{\partial\Omega_\varepsilon \cap \{r > r_0\}} g_\varepsilon\varphi \, d\sigma_x \right) dt.$$

Remark 4.3 Hereafter $\langle \cdot, \cdot \rangle_H$ is the duality pairing of a dual space H^* with H.

Remark 4.4 There are different definitions of a weak solution to problem (4.1) (e.g., [131, Sect. 3.4]). We will use several of them depending on our needs.

Similarly as, for instance, in [92, 131] we can prove that for every fixed $\varepsilon > 0$ there exists a unique weak solution to problem (4.1).

4.1.2 Auxiliary Statements

From (4.5), we deduce the inequalities

$$C_2 s^2 + k_0(0)s \le k_0(s)s \le C_3 s^2 + k_0(0)s, \tag{4.8}$$
$$|k_0(s)| \le C_4|s| + |k_0(0)| \quad \forall s \in \mathbb{R}. \tag{4.9}$$

Clearly, the same inequalities hold for the functions κ_1 and κ_2.

Lemma 4.1 *There exist positive constants C_5 and ε_0 such that for every $\varepsilon \in (0, \varepsilon_0)$ the following estimate for the weak solution u_ε to problem (4.1) holds:*

$$\max_{0 \le t \le T} \|u_\varepsilon(t, \cdot)\|_{L^2(\Omega_\varepsilon)} + \|u_\varepsilon\|_{L^2(0, T; H^1(\Omega_\varepsilon))} \le C_5 \tag{4.10}$$

Proof In the identity (4.7), we can take any $\tau \in (0, T]$ instead of T. Then, putting $\varphi = u_\varepsilon$ in (4.7), we get

$$\frac{1}{2}\|u_\varepsilon(\tau, \cdot)\|^2_{L^2(\Omega_\varepsilon)} + \int_0^\tau \left(\int_{\Omega_\varepsilon} \left(|\nabla_x u_\varepsilon|^2 + k_0(u_\varepsilon)u_\varepsilon \right) dx \right.$$

$$\left. + \varepsilon^\alpha \int_{\partial\Omega_\varepsilon^{(1)} \cap \{r > r_0\}} \kappa_1(u_\varepsilon)u_\varepsilon \, d\sigma_x + \varepsilon \int_{\partial\Omega_\varepsilon^{(2)} \cap \{r > r_0\}} \kappa_2(u_\varepsilon)u_\varepsilon \, d\sigma_x \right) dt$$

$$= \int_0^\tau \left(\int_{\Omega_\varepsilon} f_\varepsilon u_\varepsilon \, dx + \int_{\partial\Omega_0 \cap \{r < r_0\}} q_\varepsilon u_\varepsilon \, d\tilde{x} + \varepsilon^\beta \int_{\partial\Omega_\varepsilon^{(1)} \cap \{r > r_0\} \cup \partial\Omega_\varepsilon^{(2)} \cap \{r > r_0\}} g_\varepsilon u_\varepsilon \, d\sigma_x \right) dt,$$

whence using (4.8) we get

$$\frac{1}{2}\|u_\varepsilon(\tau, \cdot)\|^2_{L^2(\Omega_\varepsilon)}$$

$$+ c_0 \int_0^\tau \left(\int_{\Omega_\varepsilon} (|\nabla_x u_\varepsilon|^2 + u_\varepsilon^2) \, dx + \varepsilon^\alpha \int_{\partial\Omega_\varepsilon^{(1)} \cap \{r > r_0\}} u_\varepsilon^2 \, d\sigma_x + \varepsilon \int_{\partial\Omega_\varepsilon^{(2)} \cap \{r > r_0\}} u_\varepsilon^2 \, d\sigma_x \right) dt$$

$$\le \int_0^\tau \left(-k_0(0) \int_{\Omega_\varepsilon} u_\varepsilon \, dx - \kappa_1(0)\varepsilon^\alpha \int_{\partial\Omega_\varepsilon^{(1)} \cap \{r > r_0\}} u_\varepsilon \, d\sigma_x - \kappa_2(0)\varepsilon \int_{\partial\Omega_\varepsilon^{(2)} \cap \{r > r_0\}} u_\varepsilon \, d\sigma_x \right.$$

$$\left. + \int_{\Omega_\varepsilon} f_\varepsilon u_\varepsilon \, dx + \int_{\partial\Omega_0 \cap \{r < r_0\}} q_\varepsilon u_\varepsilon \, d\tilde{x} + + \varepsilon^\beta \int_{\partial\Omega_\varepsilon^{(1)} \cap \{r > r_0\} \cup \partial\Omega_\varepsilon^{(2)} \cap \{r > r_0\}} g_\varepsilon u_\varepsilon \, d\sigma_x \right) dt.$$

$$\tag{4.11}$$

If $\alpha \ge 1$, then with the help of Cauchy–Schwarz–Bunyakovsky inequality, (2.10) and (2.13) we derive from (4.11)

$$\frac{1}{2}\|u_\varepsilon(\tau, \cdot)\|^2_{L^2(\Omega_\varepsilon)} + c_1\|u_\varepsilon\|^2_{L^2(0, \tau; H^1(\Omega_\varepsilon))} \le c_1 \left(1 + \varepsilon^{\alpha-1} + \|f_\varepsilon\|_{L^2((0, \tau) \times \Omega_\varepsilon)} + \right.$$

$$\left. + \varepsilon^{\beta-1}\|g_\varepsilon\|_{L^2(0, \tau; H^1(\Omega^{(2)}))} + \|q_\varepsilon\|_{L^2((0, \tau) \times \partial\Omega_0 \cap \{r < r_0\})} \right) \|u_\varepsilon\|_{L^2(0, \tau; H^1(\Omega_\varepsilon))},$$

whence

$$\|u_\varepsilon\|_{L^2(0,\,\tau;H^1(\Omega_\varepsilon))} \le c_2\Big(1 + \varepsilon^{\alpha-1} + \|f_\varepsilon\|_{L^2((0,\,\tau)\times\Omega_\varepsilon)}$$
$$+ \varepsilon^{\beta-1}\|g_\varepsilon\|_{L^2(0,\,\tau;H^1(\Omega^{(2)}))} + \|q_\varepsilon\|_{L^2((0,\,\tau)\times\partial\Omega_0\cap\{r<r_0\})}\Big),$$

and

$$\max_{0\le t\le\tau}\|u_\varepsilon(t,\,\cdot)\|_{L^2(\Omega_\varepsilon)} \le c_3\Big(1 + \varepsilon^{\alpha-1} + \|f_\varepsilon\|_{L^2((0,\,\tau)\times\Omega_\varepsilon)}$$
$$+ \varepsilon^{\beta-1}\|g_\varepsilon\|_{L^2(0,\,\tau;H^1(\Omega^{(2)}))} + \|q_\varepsilon\|_{L^2((0,\,\tau)\times\partial\Omega_0\cap\{r<r_0\})}\Big).$$

Putting $\tau = T$ in those inequalities and considering (4.2)–(4.4), we get (4.10).
In the case $\alpha < 1$ with the help of (4.6), we deduce from (4.11) that

$$\frac{1}{2}\|u_\varepsilon(\tau,\,\cdot)\|^2_{L^2(\Omega_\varepsilon)} + c_4\|u_\varepsilon\|^2_{L^2(0,\,\tau;H^1(\Omega_\varepsilon))} \le c_5\Big(1 + \|f_\varepsilon\|_{L^2((0,\tau)\times\Omega_\varepsilon)}$$
$$+ \varepsilon^{\beta-1}\|g_\varepsilon\|_{L^2(0,\,\tau;H^1(\Omega^{(2)}))} + \|q_\varepsilon\|_{L^2((0,\tau)\times\partial\Omega_0\cap\{r<r_0\})}\Big)\|u_\varepsilon\|_{L^2(0,\,\tau;H^1(\Omega_\varepsilon))},$$

whence by the same way as previously we obtain (4.10).

Next we introduce the following zero-extensions

$$\widetilde{\varphi}^{(i)}(t,\,x) = \begin{cases} \varphi(t,\,x), & (t,\,x) \in (0,\,T) \times \Omega_\varepsilon^{(i)}, \\ 0, & (t,\,x) \in (0,\,T) \times (\Omega^{(i)} \setminus \Omega_\varepsilon^{(i)}), \end{cases} \quad i = 1,\,2,$$

where $\varphi : (0,\,T) \times \Omega_\varepsilon \mapsto \mathbb{R}$.
With the help of identity (2.9) similarly as in [90], we can prove the Lemma.

Lemma 4.2 *Let a sequence $\{v_\varepsilon\}_{\varepsilon>0} \subset L^2\big(0,\,T;H^1(\Omega_\varepsilon)\big)$ be uniformly bounded with respect to ε. Then*

$$\widetilde{\kappa_i(v_\varepsilon)}^{(i)} \overset{w}{\longrightarrow} \widetilde{\kappa}_i \quad \text{weakly in } L^2\big((0,\,T) \times \Omega^{(i)}\big) \text{ as } \varepsilon \to 0 \quad (i = 1,\,2)$$

and for arbitrary function $\varphi \in L^2(0,\,T;H^1(\Omega^{(i)}))$

$$\varepsilon \int_0^T \int_{\partial\Omega_\varepsilon^{(i)}\cap\{r_0<r<r_i\}} \kappa_i(v_\varepsilon)\,\varphi\,d\sigma_x dt \longrightarrow 2 \int_0^T \int_{\Omega^{(i)}} h_i^{-1}(r)\,\widetilde{\kappa}_i\,\varphi\,dxdt \quad \text{as } \varepsilon \to 0,$$

4.1.3   Convergence Theorems

Consider time-dependent multivalued functions

$$\mathbf{p}(t,\, x) = (p_0,\, p_1,\, p_2) = \begin{cases} p_0(t,\, x), & (t,\, x) \in (0,\, T) \times \Omega_0, \\ p_1(t,\, x), & (t,\, x) \in (0,\, T) \times \Omega^{(1)}, \\ p_2(t,\, x), & (t,\, x) \in (0,\, T) \times \Omega^{(2)}, \end{cases}$$

and the space $\mathbf{L}_T := L^2(0,\, T;\, \mathbf{L})$ with the scalar product

$$(\mathbf{p},\, \mathbf{q})_{\mathbf{L}_T} = \int_0^T (\mathbf{p},\, \mathbf{q})_{\mathbf{L}}\, dt,$$

(the space $\mathbf{L}$ is defined in Sect. 2.3.1). Also, we define the space

$$\widetilde{\mathbf{W}}_T := \Big\{ \mathbf{p} = (p_0,\, p_1,\, p_2) : p_0 \in L^2(0,\, T;\, H^1(\Omega_0)),\ \partial_t p_0 \in L^2(0,\, T;\, (H^1(\Omega_0))^*),$$

$$p_i \in L^2(0,\, T;\, \widetilde{H}^1(\Omega^{(i)})),\ \partial_t p_i \in L^2(0,\, T;\, (\widetilde{H}^1(\Omega^{(i)}))^*),\ i = 1,\, 2,$$

$$p_0|_{\Omega'} = p_1|_{\Omega'} = p_2|_{\Omega'} \text{ for a.e. } t \in (0,\, T) \Big\}$$

with the norm

$$\|\mathbf{p}\|_{\widetilde{\mathbf{W}}_T}^2 = \|p_0\|_{L^2(0,T;H^1(\Omega_0))}^2 + \|p_0'\|_{L^2(0,T;(H^1(\Omega_0))^*)}^2$$

$$+ \sum_{i=1}^2 \Big(\|p_i\|_{L^2(0,T;\widetilde{H}^1(\Omega^{(i)}))}^2 + \|p_i'\|_{L^2(0,T;(\widetilde{H}^1(\Omega^{(i)}))^*)}^2 \Big),$$

where $p' := \partial_t p$, $\widetilde{H}^1(\Omega^{(i)})$ is defined in Sect. 2.3.1. Obviously, the embedding $\widetilde{\mathbf{W}}_T \subset \mathbf{L}_T$ is continuous. Besides, $\widetilde{\mathbf{W}}_T \subset C([0,\, T];\, \mathbf{L})$ (see [42, Theorem 1.17]).

Theorem 4.1 *If $\alpha \geq 1$, then for the weak solution u_ε to problem (4.1) we have*

$$\left. \begin{aligned} u_\varepsilon &\xrightarrow{w} u_0 && \text{weakly in } L^2(0,\, T;\, H^1(\Omega_0)), \\ \widetilde{u}_\varepsilon^{(1)} &\xrightarrow{w} h_1 u_1 && \text{weakly in } L^2((0,\, T) \times \Omega^{(1)}), \\ \widetilde{u}_\varepsilon^{(2)} &\xrightarrow{w} h_2 u_2 && \text{weakly in } L^2((0,\, T) \times \Omega^{(2)}) \end{aligned} \right\} \quad \text{as } \varepsilon \to 0, \qquad (4.12)$$

where $\mathbf{u} := (u_0,\, u_1,\, u_2) \in \widetilde{\mathbf{W}}_T$ and it is a weak solution to the problem

$$
\begin{cases}
\partial_t u_0 - \Delta_x u_0 + k_0(u_0) = f_0, & (t, x) \in (0, T) \times \Omega_0, \\
\partial_\nu u_0 = q_0, & (t, x) \in (0, T) \times \partial\Omega_0 \cap \{r < r_0\}, \\
h_1(r)\partial_t u_1 - \operatorname{div}_{\tilde{x}}(h_1(r)\nabla_{\tilde{x}} u_1) + h_1(r)k_0(u_1) \\
\quad + 2\delta_{\alpha,1}\kappa_1(u_1) = h_1(r)f_0 + 2\delta_{\beta,1}g_0, & (t, x) \in (0, T) \times \Omega^{(1)}, \\
\partial_\nu u_1 = 0, & (t, x) \in (0, T) \times \partial\Omega^{(1)} \cap \{r = r_1\}, \\
h_2(r)\partial_t u_2 - \operatorname{div}_{\tilde{x}}(h_2(r)\nabla_{\tilde{x}} u_2) + h_2(r)k_0(u_2) \\
\quad + 2\kappa_2(u_2) = h_2(r)f_0 + 2\delta_{\beta,1}g_0, & (t, x) \in (0, T) \times \Omega^{(2)}, \\
\partial_\nu u_2 = 0, & (t, x) \in (0, T) \times \partial\Omega^{(2)} \cap \{r = r_2\}, \\
u_0 = u_1 = u_2, & (t, x) \in (0, T) \times \Omega', \\
\partial_r u_0 = \sum_{i=1}^{2} h_i(r_0)\partial_r u_i, & (t, x) \in (0, T) \times \Omega', \\
\mathbf{u}(0, x) = \mathbf{0} := (0, 0, 0).
\end{cases}
$$

$$(4.13)$$

Definition 4.2 A function $\mathbf{u} = (u_0, u_1, u_2) \in \widetilde{\mathbf{W}}_T$ is called a weak solution of problem (4.13) if $\mathbf{u}(0, x) = \mathbf{0}$ and the following integral identity takes place:

$$
\mathsf{B}_0(\mathbf{u}, \mathbf{p}; k_0(u_0), h_1 k_0(u_1), h_2 k_0(u_2), h_1\kappa_1(u_1), h_2\kappa_2(u_2)) = \mathsf{L}_0(\mathbf{p}) \quad \forall \mathbf{p} \in \widetilde{\mathbf{W}}_T,
$$

where

$$
\mathsf{B}_0(\mathbf{u}, \mathbf{p}; \ \omega_0, \widetilde{\omega}_1, \widetilde{\omega}_2, \widetilde{\kappa}_1, \widetilde{\kappa}_2)) = \int_{\Omega_0} (u_0 p_0)|_{t=T}\, dx + \sum_{i=1}^{2} \int_{\Omega^{(i)}} h_i(r)(u_i p_i)|_{t=T}\, dx
$$

$$
+ \int_0^T \bigg(-\langle \partial_t p_0, u_0 \rangle_{H^1(\Omega_0)} - \sum_{i=1}^{2} \langle h_i(r)\partial_t p_i, u_i \rangle_{\widetilde{H}^1(\Omega^{(i)})}
$$

$$
+ \int_{\Omega_0} (\nabla_x u_0 \cdot \nabla_x p_0 + \omega_0 p_0)\, dx + \sum_{i=1}^{2} \int_{\Omega^{(i)}} (h_i(r)\nabla_{\tilde{x}} u_i \cdot \nabla_{\tilde{x}} p_i + \widetilde{\omega}_i p_i)\, dx
$$

$$
+ 2\delta_{\alpha,1} \int_{\Omega^{(1)}} h_1^{-1}(r)\widetilde{\kappa}_1 p_1\, dx + 2\int_{\Omega^{(2)}} h_2^{-1}(r)\widetilde{\kappa}_2 p_2\, dx \bigg)\, dt,
$$

$$
\mathsf{L}_0(\mathbf{p}) = \int_0^T \bigg(\int_{\Omega_0} f_0 p_0\, dx + \int_{\partial\Omega_0 \cap \{r < r_0\}} q_0 p_0\, d\tilde{x} + \sum_{i=1}^{2} \int_{\Omega^{(i)}} (h_i(r)f_0 + 2\delta_{\beta,1}g_0) p_i\, dx \bigg)\, dt.
$$

Problem (4.13) is called a *homogenized problem* for problem (4.1). The existence and uniqueness of a weak solution to problem (4.13) can be proved as, e.g., in [131] using (4.5).

Proof **1**. From (4.10) and (4.9), it follows that there exists a subsequence $\{\varepsilon'\} \subset \{\varepsilon\}$ (again denoted by $\{\varepsilon\}$) such that for $i \in \{1, 2\}$ and $k \in \{1, 2, 3\}$

$$
\left.
\begin{aligned}
u_\varepsilon &\xrightarrow{w} u_0 && \text{weakly in } L^2\big(0, T; H^1(\Omega_0)\big), \\
\widetilde{u}_\varepsilon^{(i)} &\xrightarrow{w} \widetilde{u}_i := h_i u_i && \text{weakly in } L^2\big((0, T) \times \Omega^{(i)}\big), \\
\widetilde{\partial_{x_k} u_\varepsilon}^{(i)} &\xrightarrow{w} \widetilde{u}_{i,k} := h_i u_{i,k} && \text{weakly in } L^2\big((0, T) \times \Omega^{(i)}\big), \\
k_0(u_\varepsilon) &\xrightarrow{w} \omega_0 && \text{weakly in } L^2\big((0, T) \times \Omega_0\big), \\
\widetilde{k_0(u_\varepsilon)}^{(i)} &\xrightarrow{w} \widetilde{\omega}_i && \text{weakly in } L^2\big((0, T) \times \Omega^{(i)}\big), \\
\widetilde{\kappa_i(u_\varepsilon)}^{(i)} &\xrightarrow{w} \widetilde{\kappa}_i && \text{weakly in } L^2\big((0, T) \times \Omega^{(i)}\big)
\end{aligned}
\right\} \quad \text{as } \varepsilon \to 0 \quad (4.14)
$$

The estimate (4.10) also implies that for every $t \in [0, T]$ there exists a subsequence $\{\varepsilon_t\} \subset \{\varepsilon\}$ such that

$$
u_{\varepsilon_t}(t, \cdot) \xrightarrow{w} \dot{u}_0(t, \cdot) \quad \text{weakly in } L^2(\Omega_0) \tag{4.15}
$$

$$
\widetilde{u}_{\varepsilon_t}^{(i)}(t, \cdot) \xrightarrow{w} \dot{u}_i(t, \cdot) \quad \text{weakly in } L^2(\Omega^{(i)}), \ i = 1, 2. \tag{4.16}
$$

The Fubini's Theorem implies that $u_0(t, \cdot) \in L^2(\Omega_0)$ and $u_i(t, \cdot) \in L^2(\Omega^{(i)})$ for a.e. $t \in (0, T)$ $(i = 1, 2)$. Then, thanks to (4.38) and (4.16)

$$
\dot{u}_0(t, x) = u_0(t, x) \quad \text{a.e. in } (0, T) \times \Omega_0,
$$

$$
\dot{u}_i(t, x) = h_i u_i(t, x) \quad \text{a.e. in } (0, T) \times \Omega^{(i)}, \ i = 1, 2.
$$

Here unknown functions in the right-hand sides will be defined later.

2. Definition 4.1 is equivalent to the following one (cf. e.g. [131, Sect. 3.4]): a function $u_\varepsilon \in W_T(\Omega_\varepsilon)$ is called a weak solution to problem (4.1) if $u_\varepsilon(0, x) = 0$ and for all $\varphi \in H^1(\Omega_\varepsilon)$ and a.e. $t \in (0, T)$ the following integral identity holds:

$$
\langle \partial_t u_\varepsilon, \varphi \rangle_{H^1(\Omega_\varepsilon)} + \int_{\Omega_\varepsilon} \big(\nabla_x u_\varepsilon \cdot \nabla_x \varphi + k_0(u_\varepsilon)\varphi \big)\, dx
$$

$$
+ \varepsilon^\alpha \int_{\partial\Omega_\varepsilon^{(1)} \cap \{r > r_0\}} \kappa_1(u_\varepsilon)\varphi\, d\sigma_x + \varepsilon \int_{\partial\Omega_\varepsilon^{(2)} \cap \{r > r_0\}} \kappa_2(u_\varepsilon)\varphi\, d\sigma_x
$$

$$
= \int_{\Omega_\varepsilon} f_\varepsilon \varphi\, dx + \int_{\partial\Omega_0 \cap \{r < r_0\}} q_\varepsilon \varphi\, d\widetilde{x} + \varepsilon^\beta \int_{\partial\Omega_\varepsilon \cap \{r > r_0\}} g_\varepsilon \varphi\, d\sigma_x. \tag{4.17}
$$

Let us take a test function in (4.7) in the form $\varphi(t, x) = \Phi_1(x)\eta(t)$, where Φ_1 is defined in the second part of the proof of Theorem 2.2 and $\eta \in C^1([0, T])$. We get

$$
\left| \int_0^T \int_{\Omega_\varepsilon^{(1)}} \partial_{x_2} u_\varepsilon \varphi_1 \eta\, dx dt \right| = \mathcal{O}(\varepsilon), \quad \varepsilon \to 0,
$$

whence

$$\int_0^T \int_{\Omega^{(1)}} u_{1,2} \varphi_1 \eta \, dx dt = 0 \quad \forall \varphi_1 \in C_0^\infty(\Omega^{(1)}), \ \forall \eta \in C^1([0, T]).$$

Thus, $u_{1,2} = 0$ a.e. in $(0, T) \times \Omega^{(1)}$. Similarly, we show that $u_{2,2} = 0$ in $(0, T) \times \Omega^{(2)}$.

Now let us find $u_{i,1}$ and $u_{i,3}$, $i = 1, 2$. Identity (2.27) gives

$$\int_0^T \int_{\Omega^{(1)}} \widetilde{\partial_{x_m} u}_\varepsilon^{(1)} \psi \eta \, dx dt = \int_0^T \left(- \int_{\Omega^{(1)}} \widetilde{u}_\varepsilon^{(1)} (\partial_{x_m} \psi + \partial_{x_m} \ln h_1(r) \psi) \, dx \right.$$
$$\left. + \varepsilon \int_{\Omega_\varepsilon^{(1)}} Y_1 \left(\frac{x_2}{\varepsilon} \right) \partial_{x_m} \ln h_1(r) \partial_{x_2} (u_\varepsilon \psi) \, dx \right) \eta \, dt \quad (4.18)$$

for all $\psi \in C_0^\infty(\Omega^{(1)})$ and $\eta \in C^1([0, T])$; $m = 1, 3$. In the limit, (4.18) gives

$$\int_0^T \int_{\Omega^{(1)}} u_{1,m} h_1(r) \psi \eta \, dx dt = - \int_0^T \int_{\Omega^{(1)}} u_1 \partial_{x_m} (h_1(r) \psi) \eta \, dx dt,$$

whence we conclude the existence of the weak derivative $\partial_{x_m} u_1$ and $u_{1,m} = \partial_{x_m} u_1$ a.e. in $(0, T) \times \Omega^{(1)}$, $m = 1, 3$. Similarly, we show the existence of the weak derivative $\partial_{x_m} u_2$ and $u_{2,m} = \partial_{x_m} u_2$ a.e. in $(0, T) \times \Omega^{(2)}$, $m = 1, 3$.

The obtained relations provide embedding $u_i \in L^2(0, T; \widetilde{H}^1(\Omega^{(i)}))$, $i = 1, 2$.

3. Now we find conjugation conditions on the joint zone. The properties of a trace operator and (4.14) imply that for a.e. $t \in (0, T)$ the following convergence holds:

$$u_\varepsilon(t, \cdot)|_{\Omega'} \longrightarrow u_0(t, \cdot)|_{\Omega'} \quad \text{strongly in } L^2(\Omega') \text{ as } \varepsilon \to 0. \quad (4.19)$$

Identity (2.31) provides

$$-r_0^{-1} \int_0^T \int_{\Omega'} \chi_{\Omega_\varepsilon^{(1)} \cap \{r=r_0\}} u_\varepsilon \psi \eta \, d\sigma_x \, dt$$

$$= \int_0^T \left(\int_{\Omega^{(1)}} r^{-1} \left(\widetilde{\partial_r u}_\varepsilon^{(1)} \psi + \widetilde{u}_\varepsilon^{(1)} \partial_r \psi + (\ln h_1(r))' \widetilde{u}_\varepsilon^{(1)} \psi \right) dx \right.$$

$$\left. - \varepsilon \int_{\Omega_\varepsilon^{(1)}} Y_1 \left(\frac{x_2}{\varepsilon} \right) r^{-1} (\ln h_1(r))' \partial_{x_2} (u_\varepsilon \psi) \, d\sigma_x \right) \eta \, dt \quad (4.20)$$

for all $\eta \in C^1([0, T])$ and $\psi \in C^\infty(\overline{\Omega^{(1)}})$ such that $\psi|_{\partial \Omega^{(1)} \cap \{r=r_1\}} = 0$. Passing to the limit in (4.20) as $\varepsilon \to 0$, taking into account (2.16), (4.19), and (4.14), and integrating by parts, we obtain the identity

$$\int_0^T \int_{\Omega'} u_0 \psi \, \eta \, d\sigma_x dt = \int_0^T \int_{\Omega'} u_1 \, \psi \eta \, d\sigma_x dt,$$

that means $u_0|_{\Omega'} = u_1|_{\Omega'}$ for a.e. $t \in (0, T)$. Repeating the same assertions for the thin discs from the second level, we obtain

$$u_0|_{\Omega'} = u_1|_{\Omega'} = u_2|_{\Omega'} \quad \text{for a.e. } t \in (0, T). \tag{4.21}$$

4. Consider the multivalued function $\mathbf{u} = (u_0, u_1, u_2)$. On the basis of (4.14), (4.21) and the relations obtained in the item 2, we conclude that $\mathbf{u} \in \widetilde{\mathbf{W}}_T$.

Let $\eta \in C^1([0, T])$, $\varphi_0 \in C^\infty(\overline{\Omega_0})$, $\varphi_i \in C^\infty(\overline{\Omega^{(i)}})$, $i = 1, 2$, be arbitrary functions such that $\varphi_0|_{\Omega'} = \varphi_1|_{\Omega'} = \varphi_2|_{\Omega'}$, $\eta(T) = 0$. With the help of them we define

$$\Phi(t, x) = \begin{cases} \varphi_0(x)\eta(t), & (t, x) \in (0, T) \times \Omega_0, \\ \varphi_1(x)\eta(t), & (t, x) \in (0, T) \times \Omega_\varepsilon^{(1)}, \\ \varphi_2(x)\eta(t), & (t, x) \in (0, T) \times \Omega_\varepsilon^{(2)}. \end{cases}$$

Clearly, $\Phi \in W_T(\Omega_\varepsilon)$. We substitute Φ as a test function into (4.7) and utilize zero-extension operators. Then passing to the limit as $\varepsilon \to 0$ (if necessary we choose an appropriate subsequence) and bearing in mind (4.2), (4.3), (4.4), (4.9), (4.10), (2.13), (2.16), Lemmas 2.2 and 4.2, and the relations obtained above in the proof, we get

$$\mathsf{B}_0(\mathbf{u}, \mathbf{p}, \omega_0, \widetilde{\omega}_1, \widetilde{\omega}_2, \widetilde{\kappa}_1, \widetilde{\kappa}_2) = \mathsf{L}_0(\mathbf{p}), \tag{4.22}$$

where $\mathbf{p} = (\varphi_0\eta, \varphi_1\eta, \varphi_2\eta)$.

The set of multivalued functions

$$\Big\{(\varphi_0\eta, \varphi_1\eta, \varphi_2\eta) : \varphi_0 \in C^\infty(\overline{\Omega_0}), \ \varphi_1 \in C^\infty(\overline{\Omega^{(1)}}), \ \varphi_2 \in C^\infty(\overline{\Omega^{(2)}}),$$

$$\varphi_0|_{\Omega'} = \varphi_1|_{\Omega'} = \varphi_2|_{\Omega'}, \ \eta \in C^1([0, T]), \ \eta(T) = 0\Big\}$$

is dense in the space $\widetilde{\mathbf{W}}_T$ of functions $\mathbf{p}$ such that $\mathbf{p}|_{t=T} = \mathbf{0}$ (see [42, Lemma 1.12]). This means that the multivalued function $\mathbf{u} = (u_0, u_1, u_2)$ is a solution of identity (4.22) with till now unknown functions ω_0, $\widetilde{\omega}_1$, $\widetilde{\omega}_2$, $\widetilde{\kappa}_1$, and $\widetilde{\kappa}_2$.

5. In order to find them, we use Minty–Browder method. Consider the integral identity (4.7) with the test function $\varphi = u_\varepsilon$:

$$\mathsf{B}_{1,\varepsilon}(u_\varepsilon, u_\varepsilon) = \mathsf{L}_{1,\varepsilon}(u_\varepsilon).$$

Conditions (4.2), (4.3), (4.4), and relations (4.14), (2.13) and (2.16) imply that the limit of the right-hand side as $\varepsilon \to 0$ is equal to $\mathsf{L}_0(\mathbf{u})$, whence on the grounds of identity (4.22) with the test function $\mathbf{p} = \mathbf{u}$ we find that

$$\mathsf{B}_{1,\varepsilon}(u_\varepsilon, u_\varepsilon) \longrightarrow \mathsf{B}_0(\mathbf{u}, \mathbf{u}, \omega_0, \widetilde{\omega}_1, \widetilde{\omega}_2, \widetilde{\kappa}_1, \widetilde{\kappa}_2) \quad \text{as } \varepsilon \to 0. \tag{4.23}$$

Consider a monotonicity inequality

$$\frac{1}{2} \int_{\Omega_0} \left(u_\varepsilon(T, x) - p_0(T, x)\right)^2 dx + \frac{1}{2} \sum_{i=1}^{2} \int_{\Omega_\varepsilon^{(i)}} \left(u_\varepsilon(T, x) - p_i(T, x)\right)^2 dx$$

$$+ \int_0^T \Bigg(\int_{\Omega_0} |\nabla_x u_\varepsilon - \nabla_x p_0|^2 dx + \sum_{i=1}^{2} \int_{\Omega_\varepsilon^{(i)}} \left(|\nabla_{\tilde{x}} u_\varepsilon - \nabla_{\tilde{x}} p_i|^2 + (\partial_{x_2} u_\varepsilon)^2\right) dx$$

$$+ \int_{\Omega_0} \left(k_0(u_\varepsilon) - k_0(p_0)\right)(u_\varepsilon - p_0) \, dx + \sum_{i=1}^{2} \int_{\Omega_\varepsilon^{(i)}} \left(k_0(u_\varepsilon) - k_0(p_i)\right)(u_\varepsilon - p_i) \, dx$$

$$+ \varepsilon^\alpha \int_{\partial\Omega_\varepsilon^{(1)} \cap \{r > r_0\}} \left(\kappa_1(u_\varepsilon) - \kappa_1(p_1)\right)(u_\varepsilon - p_1) \, d\sigma_x$$

$$+ \varepsilon \int_{\partial\Omega_\varepsilon^{(2)} \cap \{r > r_0\}} \left(\kappa_2(u_\varepsilon) - \kappa_2(p_2)\right)(u_\varepsilon - p_2) \, d\sigma_x \Bigg) \, dt \geq 0,$$

where $\mathbf{p} = (p_0,\ p_1,\ p_2) \in \widetilde{\mathbf{W}}_T$ is arbitrary. Use zero-extensions and passing to the limit in the obtained inequality as $\varepsilon \to 0$ and utilizing (4.23), (2.16), (4.14), (2.9), and Lemma 4.2, we get

$$\frac{1}{2} \int_{\Omega_0} \left(u_0(T, x) - p_0(T, x)\right)^2 dx + \frac{1}{2} \sum_{i=1}^{2} \int_{\Omega^{(i)}} h_i(r)\left(u_i(T, x) - p_i(T, x)\right)^2 dx$$

$$+ \int_0^T \Bigg(\int_{\Omega_0} |\nabla_x u_0 - \nabla_x p_0|^2 dx + \sum_{i=1}^{2} \int_{\Omega^{(i)}} h_i(r)|\nabla_{\tilde{x}} u_i - \nabla_{\tilde{x}} p_i|^2 dx$$

$$+ \int_{\Omega_0} \left(\omega_0 - k_0(p_0)\right)(u_0 - p_0) \, dx + \sum_{i=1}^{2} \int_{\Omega^{(i)}} \left(\widetilde{\omega}_i - h_i(r)k_0(p_i)\right)(u_i - p_i) \, dx$$

$$+ 2\delta_{\alpha,1} \int_{\Omega^{(1)}} \left(h_1^{-1}(r)\widetilde{\kappa}_1 - \kappa_1(p_1)\right)(u_1 - p_1) \, dx$$

$$+ 2 \int_{\Omega^{(2)}} \left(h_2^{-1}(r)\widetilde{\kappa}_2 - \kappa_2(p_2)\right)(u_2 - p_2) \, dx \Bigg) \, dt \geq 0.$$

Setting $p_m(t, x) = u_m(t, x) - \lambda q_m(t, x)$, $m = 0, 1, 2$, in the last inequality, where $\lambda > 0$ and $\mathbf{q} = (q_0,\ q_1,\ q_2) \in \widetilde{\mathbf{W}}_T$, we obtain

$$\lambda \left\{ \frac{1}{2} \int_{\Omega_0} q_0^2 \, dx + \frac{1}{2} \sum_{i=1}^{2} \int_{\Omega^{(i)}} h_i(r) q_i^2 \, dx + \int_0^T \Bigg(\int_{\Omega_0} |\nabla_x q_0|^2 \, dx + \sum_{i=1}^{2} \int_{\Omega^{(i)}} h_i(r)|\nabla_{\tilde{x}} q_i|^2 \, dx \Bigg) dt \right\}$$

$$+ \int_0^T \Bigg(\int_{\Omega_0} \left(\omega_0 - k_0(u_0 - \lambda q_0)\right) q_0 \, dx + \sum_{i=1}^{2} \int_{\Omega^{(i)}} \left(\widetilde{\omega}_i - h_i(r)k_0(u_i - \lambda q_i)\right) q_i \, dx$$

$$+ 2\delta_{\alpha,1} \int_{\Omega^{(1)}} \left(h_1^{-1}(r)\widetilde{\kappa}_1 - \kappa_1(u_1 - \lambda q_1)\right) q_1 \, dx$$

$$+ 2 \int_{\Omega^{(2)}} \left(h_2^{-1}(r)\widetilde{\kappa}_2 - \kappa_2(u_2 - \lambda q_2)\right) q_2 \, dx \Bigg) \, dt \geq 0.$$

Sending λ to zero and taking into account continuity of k_0, κ_1, and κ_2, we get

$$\int_0^T \left(\int_{\Omega_0} (\omega_0 - k_0(u_0))q_0 \, dx + \sum_{i=1}^2 \int_{\Omega^{(i)}} (\widetilde{\omega}_i - h_i(r)k_0(u_i))q_i \, dx \right.$$
$$\left. + 2\delta_{\alpha,1} \int_{\Omega^{(1)}} (h_1^{-1}(r)\widetilde{\kappa}_1 - \kappa_1(u_1))q_1 \, dx + 2 \int_{\Omega^{(2)}} (h_2^{-1}(r)\widetilde{\kappa}_2 - \kappa_2(u_2))q_2 \, dx \right) dt \geq 0.$$

Setting $q_m := -q_m$ ($m = 0, 1, 2$), we see that in fact the equality holds. Since $\mathbf{q} = (q_0, q_1, q_2)$ is arbitrary multifunction, we deduce that

$$\omega_0 = k_0(u_0) \qquad\qquad\qquad \text{a.e. in } (0, T) \times \Omega_0,$$
$$\widetilde{\omega}_1 + 2\delta_{\alpha,1}h_1^{-1}(r)\widetilde{\kappa}_1 = h_1(r)k_0(u_1) + 2\delta_{\alpha,1}\kappa_1(u_1) \quad \text{a.e. in } (0, T) \times \Omega^{(1)},$$
$$\widetilde{\omega}_2 + 2h_2^{-1}(r)\widetilde{\kappa}_2 = h_2(r)k_0(u_2) + 2\kappa_2(u_2) \qquad \text{a.e. in } (0, T) \times \Omega^{(2)}.$$

Now it follows from (4.22) that $\mathbf{u}$ is the weak solution to problem (4.13).

6. Since all of the above assertions remain valid for arbitrary subsequence $\{\varepsilon'\}$ chosen at the beginning of the proof, the uniqueness of the weak solution to the homogenized problem (4.13) implies the limits (4.12). The theorem is proved.

Now consider the case $\alpha < 1$. Then assumption (4.6) is additionally satisfied in this case. For our investigations, we introduce the spaces

$$W_T(\Omega_0, \Omega') = \left\{ \varphi \in L^2(0, T; H^1(\Omega_0, \Omega')) : \partial_t\varphi \in L^2(0, T; (H^1(\Omega_0, \Omega'))^*) \right\},$$

and

$$\widetilde{W}_T(\Omega^{(i)}, \Omega') = \left\{ \varphi \in L^2(0, T; \widetilde{H}^1(\Omega^{(i)}, \Omega')) : \partial_t\varphi \in L^2(0, T; (\widetilde{H}^1(\Omega^{(i)}, \Omega'))^*) \right\},$$

where the spaces $H^1(\Omega_0, \Omega')$ and $\widetilde{H}^1(\Omega^{(i)}, \Omega')$ are defined in Sect. 2.3.2, $i = 1, 2$.

Theorem 4.2 *If $\alpha < 1$, then for the weak solution u_ε to problem (4.1) we have*

$$\left. \begin{array}{ll} u_\varepsilon \xrightarrow{w} u_0 & \text{weakly in } L^2(0, T; H^1(\Omega_0)), \\ \widetilde{u}_\varepsilon^{(1)} \longrightarrow 0 & \text{strongly in } L^2((0, T) \times \Omega^{(1)}), \\ \widetilde{u}_\varepsilon^{(2)} \xrightarrow{w} h_2 u_2 & \text{weakly in } L^2((0, T) \times \Omega^{(2)}) \end{array} \right\} \quad \text{as } \varepsilon \to 0, \qquad (4.24)$$

where the function $u_0 \in W_T(\Omega_0, \Omega')$ and it is a weak solution to the problem

$$\begin{cases} \partial_t u_0 - \Delta_x u_0 + k_0(u_0) = f_0, & (t, x) \in (0, T) \times \Omega_0, \\ \partial_\nu u_0 = q_0, & (t, x) \in (0, T) \times \partial\Omega_0 \cap \{r < r_0\}, \\ u_0 = 0, & (t, x) \in (0, T) \times \Omega', \\ u_0(0, x) = 0, & x \in \Omega_0, \end{cases} \qquad (4.25)$$

the function $u_2 \in \widetilde{W}_T(\Omega^{(2)}, \Omega')$ and it is a weak solution to the problem

$$
\begin{cases}
h_2(r)\partial_t u_2 - \mathrm{div}_{\tilde{x}}(h_2(r)\nabla_{\tilde{x}} u_2) + h_2(r)k_0(u_2) \\
\quad + 2\kappa_2(u_2) = h_2(r)f_0 + 2\delta_{\beta,1}g_0, & (t,\,x) \in (0,\,T) \times \Omega^{(2)}, \\
\partial_\nu u_2 = 0, & (t,\,x) \in (0,\,T) \times \partial\Omega^{(2)} \cap \{r = r_2\}, \\
u_2 = 0, & (t,\,x) \in (0,\,T) \times \Omega', \\
u_2(0,\,x) = 0, & x \in \Omega^{(2)}.
\end{cases}
$$

$$(4.26)$$

Problems (4.25) and (4.26) form the *homogenized problem* for problem (4.1).

Definition 4.3 A function $u \in L^2(0,\,T;\ H^1(\Omega_0,\,\Omega'))$ is called a weak solution to problem (4.25) if $u(0,\,x) = 0$ and

$$
\mathsf{B}_{1,0}(u,\,\varphi,\,k_0(u)) = \mathsf{L}_{1,0}(\varphi) \quad \forall\,\varphi \in W_T(\Omega_0,\,\Omega'),
$$

where

$$
\mathsf{B}_{1,0}(u,\,\varphi,\,\omega_0) = \int_{\Omega_0} (u\varphi)|_{t=T}\,dx
$$

$$
+ \int_0^T \left(- \langle \partial_t\varphi,\,u \rangle_{H^1(\Omega_0,\Omega')} + \int_{\Omega_0} (\nabla_x u \cdot \nabla_x \varphi + \omega_0 \varphi)\,dx \right)dt,
$$

$$
\mathsf{L}_{1,0}(\varphi,\,t) = \int_0^T \left(\int_{\Omega_0} f_0 \varphi\,dx + \int_{\partial\Omega_0 \cap \{r < r_0\}} q_0 \varphi\,d\tilde{x} \right)dt.
$$

Definition 4.4 A function $u \in L^2(0,\,T;\ \widetilde{H}^1(\Omega^{(2)},\,\Omega'))$ is called a weak solution to problem (4.26) if $u(0,\,x) = 0$ and

$$
\mathsf{B}_{1,2}(u,\,\varphi,\,h_2 k_0(u),\,h_2 \kappa_2(u)) = \mathsf{L}_{1,2}(\varphi) \quad \forall\,\varphi \in \widetilde{W}_T(\Omega^{(2)},\,\Omega'),
$$

where

$$
\mathsf{B}_{1,2}(u,\,\varphi,\,\tilde{\omega}_2,\,\tilde{\kappa}_2) = \int_{\Omega^{(2)}} h_2(r)(u\varphi)|_{t=T}\,dx + \int_0^T \left(- \langle h_2(r)\partial_t\varphi,\,u \rangle_{\widetilde{H}^1(\Omega^{(2)},\,\Omega')} \right.
$$

$$
\left. + \int_{\Omega^{(2)}} \left(h_2(r)\nabla_{\tilde{x}} u \cdot \nabla_{\tilde{x}} \varphi + \tilde{\omega}_2 \varphi + 2h_2^{-1}(r)\tilde{\kappa}_2 \varphi \right)dx \right)dt,
$$

$$
\mathsf{L}_{1,2}(\varphi) = \int_0^T \int_{\Omega^{(2)}} (h_2(r)f_0 + 2\delta_{\beta,1}g_0)\varphi\,dx dt.
$$

Using (4.5), we prove the existence and uniqueness of weak solutions to problems (4.25) and (4.26) as e.g. in [131].

Proof **1**. Without any changes we repeat the first item of the proof of Theorem 4.1.
  2. With the help of (4.6) and (4.8), we deduce from (4.11)

$$\frac{1}{2}\|u_\varepsilon(T,\cdot)\|^2_{L^2(\Omega_\varepsilon)}$$

$$+ c_0 \int_0^T \left(\int_{\Omega_\varepsilon} (|\nabla_x u_\varepsilon|^2 + u_\varepsilon^2)\, dx + \varepsilon^\alpha \int_{\partial\Omega_\varepsilon^{(1)} \cap \{r>r_0\}} u_\varepsilon^2\, d\sigma_x + \varepsilon \int_{\partial\Omega_\varepsilon^{(2)} \cap \{r>r_0\}} u_\varepsilon^2\, d\sigma_x \right) dt$$

$$\leq \int_0^T \left(-\omega(0) \int_{\Omega_\varepsilon} u_\varepsilon\, dx - \kappa_2(0)\varepsilon \int_{\partial\Omega_\varepsilon^{(2)} \cap \{r>r_0\}} u_\varepsilon\, d\sigma_x \right.$$

$$+ \int_{\Omega_\varepsilon} f_\varepsilon u_\varepsilon\, dx + \int_{\partial\Omega_0 \cap \{r<r_0\}} q_\varepsilon u_\varepsilon\, d\tilde{x} + \varepsilon^\beta \left. \int_{\partial\Omega_\varepsilon^{(1)} \cap \{r>r_0\} \cup \partial\Omega_\varepsilon^{(2)} \cap \{r>r_0\}} g_\varepsilon u_\varepsilon\, d\sigma_x \right) dt,$$

from where, using (4.2), (4.3), (4.4), and Lemma 4.1, it follows

$$\varepsilon \int_0^T \int_{\partial\Omega_\varepsilon^{(1)} \cap \{r>r_0\}} u_\varepsilon^2\, d\sigma_x\, dt \leq c_1 \varepsilon^{1-\alpha}. \tag{4.27}$$

Then thanks to (4.27) and Lemma 4.1 we deduce from (2.11) the estimate

$$\int_0^T \int_{\Omega_\varepsilon^{(1)}} u_\varepsilon^2\, dx dt \leq c_2 \varepsilon^{\min\{2,\,1-\alpha\}},$$

which implies that

$$\widetilde{u}_\varepsilon^{(1)} \longrightarrow 0 \quad \text{strongly in } L^2\big((0,T) \times \Omega^{(1)}\big) \text{ as } \varepsilon \to 0. \tag{4.28}$$

3. Similarly as in the second part of the proof of Theorem 4.1 we can show that $u_2 \in L^2\big(0,T;\widetilde{H}^1(\Omega^{(2)})\big)$ and

$$u_{2,2} = 0, \quad u_{2,k} = \partial_{x_k} u_2 \quad \text{a.e. in } (0,T) \times \Omega^{(2)}, \quad k = 1, 3.$$

Repeating the assertions of the third part of the proof of Theorem 4.1 and taking into account (4.28), we get

$$u_0|_{\Omega'} = u_2|_{\Omega'} = 0 \quad \text{a.e. in } (0,T) \times \Omega'.$$

4. Consider the following sets of functions:

$$S_0 = \big\{ \varphi_0 \eta : \ \varphi_0 \in C^\infty(\overline{\Omega_0}), \ \varphi_0|_{\Omega'} = 0, \ \eta \in C^1([0,T]), \ \eta(T) = 0 \big\},$$

$$S_2 = \big\{ \varphi_2 \eta : \ \varphi_2 \in C^\infty(\overline{\Omega^{(2)}}), \ \varphi_2|_{\Omega'} = 0, \ \eta \in C^1([0,T]), \ \eta(T) = 0 \big\}.$$

With the help of them, we define a function

$$\Phi(t,\,x) = \begin{cases} \varphi_0(x)\eta(t), & (t,\,x) \in (0,\,T) \times \Omega_0, \\ 0, & (t,\,x) \in (0,\,T) \times \Omega_\varepsilon^{(1)}, \\ \varphi_2(x)\eta(t), & (t,\,x) \in (0,\,T) \times \Omega_\varepsilon^{(2)}. \end{cases}$$

Clearly, $\Phi \in W_T(\Omega_\varepsilon)$. We substitute Φ as a test function into (4.7) and utilize zero-extension operators. Then passing to the limit as $\varepsilon \to 0$ (if necessary we choose an appropriate subsequence) and taking into account (2.13), (4.9), (4.2), (4.3), (4.4), (2.16), Lemma 4.2, (4.28) and relations obtained in the items **1** and **3**, we obtain

$$B_{1,0}(u_0,\,\varphi_0\eta,\,\omega_0) + B_{1,2}(u_2,\,\varphi_2\eta,\,\widetilde{\omega}_2,\,\widetilde{\kappa}_2) = L_{1,0}(\varphi_0\eta) + L_{1,2}(\varphi_2\eta).$$

Obviously, the last identity is equivalent to the following two identities:

$$B_{1,0}(u_0,\,\varphi_0\eta,\,\omega_0) = L_{1,0}(\varphi_0\eta) \quad \forall \varphi_0\eta \in S_0, \tag{4.29}$$
$$B_{1,2}(u_2,\,\varphi_2,\,\widetilde{\omega}_2,\,\widetilde{\kappa}_2) = L_{1,2}(\varphi_2) \quad \forall \varphi_2\eta \in S_2. \tag{4.30}$$

The set S_0 is dense in the space $W_T(\Omega_0,\,\Omega')$ of functions φ such that $\varphi|_{t=T} = 0$, and the set S_2 is dense in the space $\widetilde{W}_T(\Omega^{(2)},\,\Omega')$ of functions φ such that $\varphi|_{t=T} = 0$ (see [42, Lemma 1.12]). Those facts imply that u_0 is a weak solution to problem (4.25) and u_2 is a weak solution to problem (4.26) (see [131]) with still unknown functions $\omega_0,\,\widetilde{\omega}_2,\,\widetilde{\kappa}_2$.

5. In order to find them, we again use Minty–Browder method. Similarly as in the item **5** of the proof of Theorem 4.1 we find that

$$B_{1,\varepsilon}(u_\varepsilon,\,u_\varepsilon) \longrightarrow B_{1,0}(u_0,\,u_0,\,\omega_0) + B_{1,2}(u_2,\,u_2,\,\widetilde{\omega}_2,\,\widetilde{\kappa}_2) \quad \text{as } \varepsilon \to 0. \tag{4.31}$$

Consider arbitrary $\varphi_0 \in W_T(\Omega_0,\,\Omega')$, $\varphi_2 \in \widetilde{W}_T(\Omega^{(2)},\,\Omega')$, and the inequality

$$\frac{1}{2} \int_{\Omega_0} \left(u_\varepsilon(T,\,x) - \varphi_0(T,\,x)\right)^2 dx + \frac{1}{2} \int_{\Omega_\varepsilon^{(2)}} \left(u_\varepsilon(T,\,x) - \varphi_2(T,\,x)\right)^2 dx$$
$$+ \frac{1}{2} \int_{\Omega_\varepsilon^{(1)}} u_\varepsilon^2(T,\,x)\,dx + \int_0^T \left(\int_{\Omega_\varepsilon^{(1)}} |\nabla_x u_\varepsilon|^2\,dx + \int_{\Omega_\varepsilon^{(1)}} \left(k_0(u_\varepsilon) - k_0(0)\right)u_\varepsilon\,dx\right) dt$$
$$+ \int_0^T \left(\int_{\Omega_0} |\nabla_x u_\varepsilon - \nabla_x \varphi_0|^2\,dx + \int_{\Omega_\varepsilon^{(2)}} \left(|\nabla_{\widetilde{x}} u_\varepsilon - \nabla_{\widetilde{x}} \varphi_2|^2 + (\partial_{x_2} u_\varepsilon)^2\right) dx\right.$$
$$+ \int_{\Omega_0} \left(k_0(u_\varepsilon) - k_0(\varphi_0)\right)(u_\varepsilon - \varphi_0)\,dx + \int_{\Omega_\varepsilon^{(2)}} \left(k_0(u_\varepsilon) - k_0(\varphi_2)\right)(u_\varepsilon - \varphi_2)\,dx$$
$$+ \varepsilon^\alpha \int_{\partial\Omega_\varepsilon^{(1)} \cap \{r > r_0\}} \kappa_1(u_\varepsilon)u_\varepsilon\,d\sigma_x$$
$$+ \varepsilon \int_{\partial\Omega_\varepsilon^{(2)} \cap \{r > r_0\}} \left(\kappa_2(u_\varepsilon) - \kappa_2(\varphi_2)\right)(u_\varepsilon - \varphi_2)\,d\sigma_x\right) dt \geq 0.$$

Taking (4.31), (2.16), (4.14), (2.9), and Lemma 4.2 into account, we pass to the limit in the last inequality as $\varepsilon \to 0$. As a result, we get the inequality

$$\frac{1}{2} \int_{\Omega_0} \left(u_0(T, x) - \varphi_0(T, x)\right)^2 dx + \frac{1}{2} \int_{\Omega^{(2)}} h_2(r)\left(u_2(T, x) - \varphi_2(T, x)\right)^2 dx$$

$$+ \int_0^T \left(\int_{\Omega_0} |\nabla_x u_0 - \nabla_x \varphi_0|^2 dx + \int_{\Omega^{(2)}} h_2(r)|\nabla_{\tilde{x}} u_2 - \nabla_{\tilde{x}} \varphi_2|^2 dx \right.$$

$$+ \int_{\Omega_0} \left(\omega_0 - k_0(\varphi_0)\right)(u_0 - \varphi_0)\, dx + \int_{\Omega^{(2)}} \left(\widetilde{\omega}_2 - h_2(r)k_0(\varphi_2)\right)(u_2 - \varphi_2)\, dx$$

$$\left. + 2 \int_{\Omega^{(2)}} \left(h_2^{-1}(r)\widetilde{\kappa}_2 - \kappa_2(\varphi_2)\right)(u_2 - \varphi_2)\, dx \right) dt \geq 0.$$

We set $\varphi_0 = u_0 - \lambda\psi_0$, $\varphi_2 = u_2 - \lambda\psi_2$ in this inequality, where $\lambda > 0$, ψ_0, ψ_2 are arbitrary functions from $W_T(\Omega_0, \Omega')$ and $\widetilde{W}_T(\Omega^{(2)}, \Omega')$, respectively. This leads us to the inequality

$$\lambda\left\{ \frac{1}{2} \int_{\Omega_0} \psi_0^2(T, x)\, dx + \frac{1}{2} \int_{\Omega^{(2)}} h_2(r)\psi_2^2(T, x)\, dx + \int_0^T \left(\int_{\Omega_0} |\nabla_x \psi_0|^2 dx \right.\right.$$

$$\left.\left. + \int_{\Omega^{(2)}} h_2(r)|\nabla_{\tilde{x}} \psi_2|^2 dx \right) dt \right\}$$

$$+ \int_0^T \left(\int_{\Omega_0} \left(\omega_0 - k_0(u_0 - \lambda\psi_0)\right)\psi_0\, dx + \int_{\Omega^{(2)}} \left(\widetilde{\omega}_2 - h_2(r)k_0(u_2 - \lambda\psi_2)\right)\psi_2\, dx$$

$$+ 2 \int_{\Omega^{(2)}} \left(h_2^{-1}(r)\widetilde{\kappa}_2 - \kappa_2(u_2 - \lambda\psi_2)\right)\psi_2\, dx \right) dt \geq 0.$$

Now we pass to the limit as $\lambda \to 0$ taking into account continuity of k_0 and κ_2:

$$\int_0^T \left(\int_{\Omega_0} \left(\omega_0 - k_0(u_0)\right)\psi_0\, dx + \int_{\Omega^{(2)}} \left(\widetilde{\omega}_2 - h_2(r)k_0(u_2)\right)\psi_2\, dx$$

$$+ 2 \int_{\Omega^{(2)}} \left(h_2^{-1}(r)\widetilde{\kappa}_2 - \kappa_2(u_2)\right)\psi_2\, dx \right) dt \geq 0.$$

Substituting $\psi_0 := -\psi_0$, $\psi_2 := -\psi_2$, we see that in fact the equality holds. Since $\psi_0 \in W_T(\Omega_0, \Omega')$ and $\psi_2 \in \widetilde{W}_T(\Omega^{(2)}, \Omega')$ are arbitrary functions, we get

$$\begin{aligned} \omega_0 &= k_0(u_0) && \text{a.e. in } (0,\ T) \times \Omega_0, \\ \widetilde{\omega}_2 + 2h_2^{-1}(r)\widetilde{\kappa}_2 &= h_2(r)k_0(u_2) + 2\kappa_2(u_2) && \text{a.e. in } (0,\ T) \times \Omega^{(2)}. \end{aligned} \tag{4.32}$$

Thus, the first relation in (4.32) and identity (4.29) imply that the function u_0 is the weak solution to problem (4.25), and the second relation in (4.32) and identity (4.30) imply that u_2 is the weak solution of problem (4.26).

6. Since all of the above assertions remain valid for arbitrary subsequence $\{\varepsilon'\}$ chosen at the beginning of the proof, the uniqueness of weak solutions to problems (4.25) and (4.26) implies that relations (4.24) hold. The theorem is proved.

4.2 Problem with Alternating Robin and Dirichlet Boundary Conditions

4.2.1 Statement of the Problem

Consider the following semilinear parabolic problem:

$$\begin{cases} \partial_t v_\varepsilon(t, x) - \Delta_x v_\varepsilon(t, x) + k_0(v_\varepsilon) = f_\varepsilon, & (t, x) \in (0, T) \times \Omega_\varepsilon, \\ \partial_\nu v_\varepsilon(t, x) + \varepsilon^\alpha \kappa_1(v_\varepsilon) = \varepsilon^\beta g_\varepsilon(t, x), & (t, x) \in (0, T) \times \partial\Omega_\varepsilon^{(1)} \cap \{r > r_0\}, \\ v_\varepsilon(t, x) = 0, & (t, x) \in (0, T) \times \partial\Omega_\varepsilon^{(2)} \cap \{r > r_0\}, \\ \partial_\nu v_\varepsilon(t, x) = 0, & (t, x) \in (0, T) \times \partial\Omega_\varepsilon \cap \{r = r_0\}, \\ \partial_\nu v_\varepsilon(t, x) = q_\varepsilon(t, x), & (t, x) \in (0, T) \times \partial\Omega_0 \cap \{r < r_0\}, \\ v_\varepsilon(0, x) = 0, & x \in \Omega_\varepsilon. \end{cases}$$

(4.33)

Here the functions $f_\varepsilon, g_\varepsilon, q_\varepsilon, k_0, \kappa_1$ satisfy the same conditions as in Sect. 4.1.1. We introduce the space

$$W_T(\Omega_\varepsilon, \partial\Omega_\varepsilon^{(2)} \cap \{r > r_0\}) := \big\{\varphi \in L^2(0, T; H^1(\Omega_\varepsilon, \partial\Omega_\varepsilon^{(2)} \cap \{r > r_0\})) :$$
$$\partial_t \varphi \in L^2(0, T; (H^1(\Omega_\varepsilon, \partial\Omega_\varepsilon^{(2)} \cap \{r > r_0\}))^*)\big\}$$

where the space $H^1(\Omega_\varepsilon, \partial\Omega_\varepsilon^{(2)} \cap \{r > r_0\})$ is defined in Sect. 2.1. It is known (see e.g., [42, Theorem 1.17]) that $W_T(\Omega_\varepsilon, \partial\Omega_\varepsilon^{(2)} \cap \{r > r_0\}) \subset C([0, T]; L^2(\Omega_\varepsilon))$.

Definition 4.5 A function $v_\varepsilon \in W_T(\Omega_\varepsilon, \partial\Omega_\varepsilon^{(2)} \cap \{r > r_0\})$ is called a weak solution to problem (4.33) if $v_\varepsilon(0, x) = 0$ and the integral identity

$$\mathsf{B}_{2,\varepsilon}(v_\varepsilon, \varphi) = \mathsf{L}_{2,\varepsilon}(\varphi) \tag{4.34}$$

holds for any function $\varphi \in W_T(\Omega_\varepsilon, \partial\Omega_\varepsilon^{(2)} \cap \{r > r_0\})$, where

$$B_{2,\varepsilon}(v_\varepsilon, \varphi) = \int_{\Omega_\varepsilon} (v_\varepsilon\varphi)|_{t=T}\, dx + \int_0^T \bigg(-\langle\partial_t\varphi, v_\varepsilon\rangle_{H^1(\Omega_\varepsilon,\, \partial\Omega_\varepsilon^{(2)}\cap\{r>r_0\})}$$

$$+ \int_{\Omega_\varepsilon} (\nabla_x v_\varepsilon \cdot \nabla_x\varphi + \omega(v_\varepsilon)\varphi)\, dx + \varepsilon^\alpha \int_{\partial\Omega_\varepsilon^{(1)}\cap\{r>r_0\}} \kappa_1(v_\varepsilon)\varphi\, d\sigma_x \bigg)\, dt,$$

$$L_{2,\varepsilon}(\varphi) = \int_0^T \bigg(\int_{\Omega_\varepsilon} f_\varepsilon\varphi\, dx + \int_{\partial\Omega_0\cap\{r<r_0\}} q_\varepsilon\varphi\, d\tilde{x} + \varepsilon^\beta \int_{\partial\Omega_\varepsilon^{(1)}\cap\{r>r_0\}} g_\varepsilon\varphi\, d\sigma_x \bigg)\, dt.$$

Using (4.5), we prove the existence and uniqueness of the weak solution to problem (4.33) for each fixed $\varepsilon > 0$ (see e.g. [131]).

4.2.2 Convergence Theorems

With the help of (4.8) similarly as in Sect. 4.1.2, we prove a priory estimate.

Lemma 4.3 *There exist positive constants C_0 and ε_0 such that for every $\varepsilon \in (0, \varepsilon_0)$ the following estimate for the weak solution v_ε to problem (4.33) holds:*

$$\max_{0\le t\le T} \|v_\varepsilon(t, \cdot)\|_{L^2(\Omega_\varepsilon)} + \|v_\varepsilon\|_{L^2(0,\, T;\, H^1(\Omega_\varepsilon))} \le C_0.$$

Theorem 4.3 *If $\alpha \ge 1$, then the solution v_ε to problem (4.33) satisfies the relations*

$$\left.\begin{array}{l} v_\varepsilon \xrightarrow{w} v_0 \quad \text{weakly in } L^2(0,\, T;\, H^1(\Omega_0)), \\[4pt] \widetilde{v}_\varepsilon^{(1)} \xrightarrow{w} h_1 v_1 \text{ weakly in } L^2((0,\, T)\times\Omega^{(1)}), \\[4pt] \widetilde{v}_\varepsilon^{(2)} \xrightarrow{w} 0 \quad \text{weakly in } L^2(0,\, T;\, H^1(\Omega^{(2)})) \end{array}\right\} \quad \text{as } \varepsilon\to 0, \qquad (4.35)$$

where $v_0 \in W_T(\Omega_0, \Omega')$ and it is a weak solution to problem (4.25), $v_1 \in \widetilde{W}_T(\Omega^{(1)}, \Omega')$ and it is a weak solution to the problem

$$\begin{cases} h_1(r)\partial_t v_1 - \operatorname{div}_{\tilde{x}}(h_1(r)\nabla_{\tilde{x}} v_1) + h_1(r)k_0(v_1) & \\ \quad + 2\delta_{\alpha,1}\kappa_1(v_1) = h_1(r)f_0 + 2\delta_{\beta,1}g_0, & (t, x) \in (0, T)\times\Omega^{(1)}, \\ \partial_\nu v_1 = 0, & (t, x) \in (0, T)\times\partial\Omega^{(1)}\cap\{r = r_1\}, \\ v_1 = 0, & (t, x) \in (0, T)\times\Omega', \\ v_1(0, x) = 0, & x \in \Omega^{(1)}. \end{cases}$$

$$(4.36)$$

Definition 4.6 A function $v \in \widetilde{W}_T(\Omega^{(1)}, \Omega')$ is called a weak solution to problem (4.36) if it satisfies the initial condition $v(0, x) = 0$ and the integral identity

$$B_{2,1}(v, \varphi, h_1 k_0(v), h_1\kappa_1(v)) = L_{2,1}(\varphi) \quad \forall\varphi \in \widetilde{W}_T(\Omega^{(1)}, \Omega'),$$

where

$$\mathsf{B}_{2,1}(v,\ \varphi,\ \tilde{\omega}_1,\ \tilde{\kappa}_1) = \int_{\Omega^{(1)}} h_1(r)(v\varphi)|_{t=T}\,dx + \int_0^T \Big(-\langle \partial_t\varphi,\ h_1(r)v\rangle_{\tilde{H}^1(\Omega^{(1)},\,\Omega')}$$

$$+ \int_{\Omega^{(1)}} \big(h_1(r)\nabla_{\tilde{x}}v\cdot\nabla_{\tilde{x}}\varphi + \tilde{\omega}_1\varphi + 2\delta_{\alpha,1}h_1^{-1}(r)\tilde{\kappa}_1\varphi\big)\,dx\Big)\,dt,$$

$$\mathsf{L}_{2,1}(\varphi) = \int_0^T \int_{\Omega^{(1)}} (h_1(r)f_0 + 2\delta_{\beta,1}g_0)\varphi\,dxdt.$$

Due to the assumptions for the functions f_0, g_0, k_0, and κ_1, there exists a unique weak solution to problem (4.36) (see e.g., [131]).

Proof **1.** Note that for any function $\varphi \in L^2(0,\ T;\ H^1(\Omega_\varepsilon^{(2)},\ \partial\Omega_\varepsilon^{(2)}\cap\{r > r_0\}))$ its zero-extension $\tilde{\varphi}^{(2)}$ belongs to the space $L^2(0,\ T;\ H^1(\Omega^{(2)}))$. Inequality (2.12) and Lemma 4.3 imply that

$$\|\tilde{v}_\varepsilon^{(2)}\|_{L^2((0,\,T)\times\Omega^{(2)})} \le c_0\varepsilon\|\nabla_x v_\varepsilon\|_{L^2((0,\,T)\times\Omega_\varepsilon^{(2)})} \le c_1\varepsilon.$$

Therefore, the last limit in (4.35) holds.

2. Lemma 4.3 and inequality (4.9) imply that there exists a subsequence $\{\varepsilon'\} \subset \{\varepsilon\}$ (again denoted by $\{\varepsilon\}$) such that the following convergence holds:

$$\left.\begin{aligned}
v_\varepsilon &\xrightarrow{w} v_0 && \text{weakly in } L^2(0,\ T;\ H^1(\Omega_0)),\\
\tilde{v}_\varepsilon^{(1)} &\xrightarrow{w} \tilde{v}_1 := h_1 v_1 && \text{weakly in } L^2((0,\ T)\times\Omega^{(1)}),\\
\widetilde{\partial_{x_k}v_\varepsilon}^{(1)} &\xrightarrow{w} \tilde{v}_{1,k} := h_1 v_{1,k} && \text{weakly in } L^2((0,\ T)\times\Omega^{(1)}),\\
k_0(v_\varepsilon) &\xrightarrow{w} \omega_0 && \text{weakly in } L^2((0,\ T)\times\Omega_0),\\
\widetilde{k_0(v_\varepsilon)}^{(1)} &\xrightarrow{w} \tilde{\omega}_1 && \text{weakly in } L^2((0,\ T)\times\Omega^{(1)}),\\
\widetilde{\kappa_1(v_\varepsilon)}^{(1)} &\xrightarrow{w} \tilde{\kappa}_1 && \text{weakly in } L^2((0,\ T)\times\Omega^{(1)}),
\end{aligned}\right\} \tag{4.37}$$

as $\varepsilon \to 0$ ($k = 1,\ 2,\ 3$).

Similar as in the proof of Theorem 4.1 we show that for every $t \in [0, T]$, there exists a subsequence $\{\varepsilon_t\} \subset \{\varepsilon\}$ such that

$$\left.\begin{aligned}
v_{\varepsilon_t}(t,\ \cdot) &\xrightarrow{w} v_0(t,\ \cdot) && \text{weakly in } L^2(\Omega_0),\\
\tilde{v}_{\varepsilon_t}^{(i)}(t,\ \cdot) &\xrightarrow{w} h_1 v_1(t,\ \cdot) && \text{weakly in } L^2(\Omega^{(1)}),
\end{aligned}\right\} \quad \text{as } \varepsilon \to 0, \tag{4.38}$$

and

- $v_{1,2} = 0$ a.e. in $(0,\ T)\times\Omega^{(1)}$;
- there exist weak derivatives $\partial_{x_k}v_1 = v_{1,k}$ a.e. in $(0,\ T)\times\Omega^{(1)}$, $k = 1,\ 3$;
- $v_0|_{\Omega'} = v_1|_{\Omega'} = 0$ for a.e. $(t,\ x) \in (0,\ T)\times\Omega'$.

3. Let us define a function

$$\Phi(t, x) = \begin{cases} \varphi_0(x)\eta(t), & (t, x) \in (0, T) \times \Omega_0, \\ \varphi_1(x)\eta(t), & (t, x) \in (0, T) \times \Omega_\varepsilon^{(1)}, \\ 0, & (t, x) \in (0, T) \times \Omega_\varepsilon^{(2)}, \end{cases}$$

where $\varphi_0\eta \in S_0$ (it is defined in the item **4** of the proof of Theorem 4.1) and $\varphi_1\eta$ belongs to the function set

$$S_1 = \left\{ \varphi_1\eta : \; \varphi_1 \in C^\infty(\overline{\Omega^{(1)}}), \; \varphi_1|_{\Omega'} = 0, \; \eta \in C^1([0, T]), \; \eta(T) = 0 \right\}.$$

Clearly, $\Phi \in W_T(\Omega_\varepsilon, \partial\Omega_\varepsilon^{(2)} \cap \{r > r_0\})$. Next we substitute Φ as a test function into (4.34) and utilize the zero-extension operators. Bearing in mind (2.13), (4.9), (4.2), (4.3), (4.4), (2.16), Lemma 4.2, and the relations, obtained in the previous parts of the proof, we pass to the limit as $\varepsilon \to 0$. As a result, we obtain

$$B_{1,0}(v_0, \varphi_0\eta, \omega_0) + B_{2,1}(v_1, \varphi_1\eta, \widetilde{\omega}_1, \widetilde{\kappa}_1) = L_{1,0}(\varphi_0\eta) + L_{2,1}(\varphi_1\eta).$$

Obviously, the last identity is equivalent to the following ones:

$$B_{1,0}(v_0, \varphi_0\eta, \omega_0) = L_{1,0}(\varphi_0\eta) \quad \forall \varphi_0\eta \in S_0, \tag{4.39}$$

$$B_{2,1}(v_1, \varphi_1\eta, \widetilde{\omega}_1, \widetilde{\kappa}_1) = L_{2,1}(\varphi_1\eta) \quad \forall \varphi_1\eta \in S_1. \tag{4.40}$$

The set S_0 is dense in a set of functions $\varphi \in W_T(\Omega_0, \Omega')$ such that $\varphi|_{t=T} = 0$, and the set S_1 is dense in a set of functions $\varphi \in \widetilde{W}_T(\Omega^{(1)}, \Omega')$ such that $\varphi|_{t=T} = 0$ (see [42, Lemma 1.12]). Thus, identity (4.39) implies that $v_0 \in W_T(\Omega_0, \Omega')$ is a weak solution to problem (4.25) with still unknown function ω_0, and identity (4.40) provides that $v_1 \in \widetilde{W}_T(\Omega^{(1)}, \Omega')$ is a weak solution to problem (4.36) with unknown functions $\widetilde{\omega}_1$ and $\widetilde{\kappa}_1$.

4. Similarly as in the item **5** of the proof of Theorem 4.1 with the help of the inequality

$$\frac{1}{2} \int_{\Omega_0} \left(v_\varepsilon(T, x) - \varphi_0(T, x) \right)^2 dx + \frac{1}{2} \int_{\Omega_\varepsilon^{(1)}} \left(v_\varepsilon(T, x) - \varphi_1(T, x) \right)^2 dx$$

$$+ \frac{1}{2} \int_{\Omega_\varepsilon^{(2)}} \left(v_\varepsilon(T, x) \right)^2 dx + \int_{\Omega_\varepsilon^{(2)}} |\nabla_x v_\varepsilon|^2 dx$$

$$+ \int_0^T \Big(\int_{\Omega_0} |\nabla_x v_\varepsilon - \nabla_x \varphi_0|^2 dx + \int_{\Omega_\varepsilon^{(1)}} \left(|\nabla_{\widetilde{x}} v_\varepsilon - \nabla_{\widetilde{x}} \varphi_1|^2 + (\partial_{x_2} v_\varepsilon)^2 \right) dx$$

$$+ \int_{\Omega_0} \left(k_0(v_\varepsilon) - k_0(\varphi_0) \right)(v_\varepsilon - \varphi_0) \, dx + \int_{\Omega_\varepsilon^{(1)}} \left(k_0(v_\varepsilon) - k_0(\varphi_1) \right)(v_\varepsilon - \varphi_1) \, dx$$

$$+ \int_{\Omega_\varepsilon^{(2)}} \left(k_0(v_\varepsilon) - k_0(0) \right)v_\varepsilon \, dx + \varepsilon^\alpha \int_{\partial\Omega_\varepsilon^{(1)} \cap \{r > r_0\}} \left(\kappa_1(v_\varepsilon) - \kappa_1(\varphi_1) \right)(v_\varepsilon - \varphi_1) \, d\sigma_x \Big) dt \geq 0$$

(here φ_0 and φ_1 are arbitrary functions from the spaces $W_T(\Omega_0, \Omega')$ and $\widetilde{W}_T(\Omega^{(1)}, \Omega')$, respectively) we show that

$$\omega_0 = k_0(v_0) \quad \text{a.e. in } (0, T) \times \Omega_0$$
$$\widetilde{\omega}_1 + 2\delta_{\alpha,1} h_1^{-1}(r)\widetilde{\kappa}_1 = h_1(r)k_0(v_1) + 2\delta_{\alpha,1}\kappa_1(v_1) \quad \text{a.e. in } (0, T) \times \Omega^{(1)}.$$

Thus, v_0 and v_1 are weak solutions to problems (4.25) and (4.36), respectively.

5. Since all of the above assertions remain valid for arbitrary subsequence $\{\varepsilon'\}$ chosen at the beginning of the proof, the uniqueness of weak solutions to problems (4.25) and (4.36) implies that relations (4.35) hold. The theorem is proved.

Now consider the case $\alpha < 1$. Then assumption (4.6) is satisfied in this case.

Theorem 4.4 *If $\alpha < 1$, then for the solution v_ε to problem (4.33) we have*

$$\left. \begin{array}{l} v_\varepsilon \xrightarrow{w} v_0 \text{ weakly in } L^2\big(0, T; H^1(\Omega_0)\big), \\ \widetilde{v}_\varepsilon^{(1)} \longrightarrow 0 \text{ strongly in } L^2\big((0, T) \times \Omega^{(1)}\big), \\ \widetilde{v}_\varepsilon^{(2)} \xrightarrow{w} 0 \text{ weakly in } L^2\big(0, T; H^1(\Omega^{(2)})\big) \end{array} \right\} \quad \text{as } \varepsilon \to 0, \qquad (4.41)$$

where $v_0 \in L^2\big(0, T; H^1(\Omega_0, \Omega')\big)$ and it is a weak solution to problem (4.25).

Proof We repeat the items **1** and **2** of the proof of Theorem 4.3. Also repeating assertions of the item **2** of the proof of Theorem 4.2, we can prove that

$$\widetilde{v}_\varepsilon^{(1)} \longrightarrow 0 \quad \text{strongly in } L^2\big((0, T) \times \Omega^{(1)}\big) \text{ as } \varepsilon \to 0.$$

The first limit in (4.37) and compactness of the embedding $L^2\big(0, T; H^1(\Omega_0)\big) \subset L^2\big((0, T) \times \Omega_0\big)$ imply existence of a subsequence $\{\varepsilon'\} \subset \{\varepsilon\}$ (again denoted by $\{\varepsilon\}$) such that

$$v_\varepsilon \to v_0 \quad \text{strongly in } L^2\big((0, T) \times \Omega_0\big) \text{ and a.e. in } (0, T) \times \Omega_0.$$

By virtue of the continuity of k_0, we have that

$$k_0(v_\varepsilon) \longrightarrow k_0(v_0) \quad \text{a.e. in } (0, T) \times \Omega_0 \text{ as } \varepsilon \to 0.$$

Using [70, Lemma 1.3] and Lemma 4.3, we get $\omega_0 = k_0(v_0)$ a.e. in $(0, T) \times \Omega_0$.

Let us define a function

$$\Phi(t, x) := \begin{cases} \varphi_0(x)\eta(t), & (t, x) \in (0, T) \times \Omega_0, \\ 0, & (t, x) \in (0, T) \times \Omega_\varepsilon^{(1)} \cup \Omega_\varepsilon^{(2)}, \end{cases}$$

where $\varphi_0 \eta \in S_0$ (it is defined in the item **4** of the proof of Theorem 4.1). Clearly, $\Phi \in W_T(\Omega_\varepsilon, \partial\Omega_\varepsilon^{(2)} \cap \{r > r_0\})$. We substitute Φ into integral identity (4.34) as a test function. Passing to the limit and taking into account the obtained relations, we get the integral identity (4.39) for the function v_0. Since S_0 is dense in a set of functions $\varphi \in W_T(\Omega_0, \Omega')$ such that $\varphi|_{t=T} = 0$, v_0 is a weak solution to problem (4.25) (see Definition 4.3).

The standard remark on the uniqueness of the weak solution to problem (4.25) completes the proof of the theorem.

4.3   Conclusions to this Chapter

The results obtained in this chapter show that the influence of the geometric structure of the thick junction Ω_ε and boundary conditions on the asymptotic behavior of the solutions to the semilinear parabolic problems remains the same as for the elliptic problems considered in Chap. 2 in the case if the parameter $\alpha \geq 1$ (see the conclusions to Chap. 2).

Indeed, in the case $\alpha \geq 1$, the homogenized problem for problem (4.1) is a non-standard problem in the anisotropic Sobolev space $\widetilde{\mathbf{W}}_T$ of multivalued functions, and the nonlinear Robin boundary conditions are transformed (as $\varepsilon \to 0$) by the same way into two summands of the corresponding differential equations in the domains $\Omega^{(i)}$, $i = 1, 2$.

The case $\alpha < 1$, which is not considered in Chap. 2, is qualitatively quite different. The initial problem (4.1) is divided (as $\varepsilon \to 0$) into two independent problems in the domains Ω_0 and $\Omega^{(2)}$.

If we interpret problem (4.1) as a mathematical model of the heat radiation, then conditions $\alpha < 1$ and $\kappa_1(0) = 0$ mean that there is an intensive heat exchange on the surfaces of the thin discs from the first level with cold environment. As a result, the thin discs from the first level are quickly cooled down, cooling at the same time the junction's body (as a result, we have the homogeneous Dirichlet condition on Ω' in problems (4.25) and (4.26)).

Similar conclusions are valid for problem (4.33). In the case $\alpha \geq 1$, the initial problem is divided (as $\varepsilon \to 0$) into two independent problems thanks to the homogeneous Dirichlet boundary conditions on the surfaces of the thin discs from the second level. If $\alpha < 1$, the solution to problem (4.33) tends to zero in the thin discs from both levels.

Chapter 5
Asymptotic Approximations for Solutions to Semilinear Elliptic and Parabolic Problems

Here another approach to study the asymptotic behavior of solutions to BVPs in thick two-level junctions of the type 3:2:2 is demonstrated. We consider two semilinear problems (elliptic and parabolic) in the thick junction Ω_ε described in Sect. 2.1. Nonlinear perturbed Robin boundary conditions are imposed on the surfaces of the thin discs from both levels in each of the problems. Approximations for the solutions to those problems are constructed and asymptotic estimates in Sobolev spaces are proved. Two-scale asymptotic expansion method and the method of matching asymptotic expansions are used.

5.1 Semilinear Elliptic Problem

5.1.1 Statement of the Problem

In Ω_ε, we consider the semilinear boundary-value problem

$$
\begin{cases}
-\Delta_x u_\varepsilon(x) + k_0(u_\varepsilon(x)) = f_\varepsilon(x), & x \in \Omega_\varepsilon, \\
\partial_\nu u_\varepsilon(x) + \varepsilon^\alpha \kappa_1(u_\varepsilon(x)) = \varepsilon^\beta g_\varepsilon(x), & x \in \partial\Omega_\varepsilon^{(1)} \cap \{r > r_0\}, \\
\partial_\nu u_\varepsilon(x) + \varepsilon \kappa_2(u_\varepsilon(x)) = \varepsilon^\beta g_\varepsilon(x), & x \in \partial\Omega_\varepsilon^{(2)} \cap \{r_0 < r < r_2\}, \\
\partial_\nu u_\varepsilon(x) + \kappa_2(u_\varepsilon(x)) = 0, & x \in \partial\Omega_\varepsilon^{(2)} \cap \{r = r_2\}, \\
\partial_\nu u_\varepsilon(x) = 0, & x \in \partial\Omega_\varepsilon \cap \{r = r_0\}, \\
\partial_{x_2}^p u_\varepsilon(x_1, 0, x_3) = \partial_{x_2}^p u_\varepsilon(x_1, l, x_3), & p = 0, 1, \ x \in \partial\Omega_0 \cap \{r < r_0\}.
\end{cases}
\tag{5.1}
$$

Here $\alpha \geq 1$, $\beta \geq 1$ are parameters; the functions k_0, κ_1 and κ_2 satisfy assumptions (4.5); $f_\varepsilon \in L^2(\Omega_\varepsilon)$, $g_\varepsilon \in H^1(\Omega^{(2)})$, and

© The Author(s), under exclusive license to Springer Nature Switzerland AG 2019
T. Mel'nyk and D. Sadovyi, *Multiple-Scale Analysis of Boundary-Value Problems in Thick Multi-Level Junctions of Type 3:2:2*, SpringerBriefs in Mathematics, https://doi.org/10.1007/978-3-030-35537-1_5

$$\|g_\varepsilon\|_{L^2(\Omega^{(2)})} + \|\partial_{x_2} g_\varepsilon\|_{L^2(\Omega^{(2)})} \leq C_0. \tag{5.2}$$

The thick junction Ω_ε is described in Sect. 2.1. In this chapter, we additionally assume that the functions h_1, h_2 describing the geometric structure of the thin discs are constant in a neighborhood of r_0, i.e., there exists a constant $\delta_0 > 0$ such that $h_i(r) = h_i(r_0)$ for all $r \in [r_0, r_0 + \delta_0]$, $i = 1, 2$.

Definition 5.1 A function

$$u_\varepsilon \in H^1_{\mathrm{per}}(\Omega_\varepsilon) := \{\varphi \in H^1(\Omega_\varepsilon) : \varphi(x_1, 0, x_3) = \varphi(x_1, l, x_3), \ r < r_0\}$$

is called a weak solution to problem (5.1) if for all $\varphi \in H^1_{\mathrm{per}}(\Omega_\varepsilon)$ the identity

$$\int_{\Omega_\varepsilon} (\nabla_x u_\varepsilon \cdot \nabla_x \varphi + k_0(u_\varepsilon)\varphi) \, dx + \varepsilon^\alpha \int_{\partial\Omega_\varepsilon^{(1)} \cap \{r > r_0\}} \kappa_1(u_\varepsilon)\varphi \, d\sigma_x$$

$$+ \varepsilon \int_{\partial\Omega_\varepsilon^{(2)} \cap \{r_0 < r < r_2\}} \kappa_2(u_\varepsilon)\varphi \, d\sigma_x + \int_{\partial\Omega_\varepsilon^{(2)} \cap \{r = r_2\}} \kappa_2(u_\varepsilon)\varphi \, d\sigma_x = \mathsf{L}_\varepsilon(\varphi) \tag{5.3}$$

holds. Here

$$\mathsf{L}_\varepsilon(\varphi) := \int_{\Omega_\varepsilon} f_\varepsilon \varphi \, dx + \varepsilon^\beta \int_{\partial\Omega_\varepsilon^{(1)} \cap \{r > r_0\} \cup \partial\Omega_\varepsilon^{(2)} \cap \{r_0 < r < r_2\}} g_\varepsilon \varphi \, d\sigma_x. \tag{5.4}$$

Similarly as in [90], we can prove that for every fixed $\varepsilon > 0$ there exists a unique weak solution to problem (5.1).

5.1.2 Formal Asymptotic Expansions

Only in this section, we assume that the functions f_ε, g_ε are independent of ε, i.e., $f_\varepsilon = f_0$ in $\Omega_0 \cup \Omega^{(2)}$ and $g_\varepsilon = g_0$ in $\Omega^{(2)}$, and they are smooth enough.

Outer Expansions

We propose the following asymptotic ansatzes for the solution u_ε:

$$u_\varepsilon(x) \approx u_0^+(x) + \varepsilon u_1^+(x) + \ldots, \quad \text{in } \Omega_0, \tag{5.5}$$

and

$$u_\varepsilon(x) \approx u_0^{i,-}(x) + \varepsilon u_1^{i,-}\left(x, \frac{x_2}{\varepsilon} - j\right) + \varepsilon^2 u_2^{i,-}\left(x, \frac{x_2}{\varepsilon} - j\right) + \ldots, \quad \text{in } \Omega_\varepsilon^{(i)}(j), \tag{5.6}$$

where $j \in \{0, 1, \ldots, N - 1\}$, $i = 1, 2$. Expansions (5.5) and (5.6) are usually called the *outer expansions*.

Using Taylor's formula, we get

$$k_0(u_\varepsilon(x)) = k_0(u_0^+(x)) + \mathcal{O}(\varepsilon), \quad \varepsilon \to 0 \quad (x \in \Omega_0). \tag{5.7}$$

Substituting (5.5) into the first equation to problem (5.1) and into the boundary conditions on $\partial\Omega_0 \cap \{r < r_0\}$, utilizing (5.7), and collecting coefficients at the same powers of ε, we obtain the following relations for the function u_0^+:

$$\begin{cases} -\Delta_x u_0^+ + k_0(u_0^+) = f_0, & x \in \Omega_0, \\ \partial_{x_2}^p u_0^+(x_1, 0, x_3) = \partial_{x_2}^p u_0^+(x_1, l, x_3), & x \in \partial\Omega_0 \cap \{r < r_0\}, \quad p = 0, 1. \end{cases}$$

Now let us find relations for the main terms of the outer expansions in the domains $\Omega_\varepsilon^{(i)}$, $i = 1, 2$. Considering the functions $u_k^{i,-}(x, \xi_2 - j)$ to be smooth, we write down their Taylor series with respect to the "slow" variable x_2 in a neighborhood of $x_2 = \varepsilon(j + l_i)$ for fixed $j \in \{0, 1, \dots, N - 1\}$ and $i = 1, 2$. Then (5.6) reads

$$u_\varepsilon(x) \approx u_0^{i,-}(x_1, \varepsilon(j + l_i), x_3) + \sum_{k=1}^{\infty} \varepsilon^k U_k^{i,-}\left(x_1, \varepsilon(j + l_i), x_3, \frac{x_2}{\varepsilon} - j\right), \quad x \in \Omega_\varepsilon^{(i)}(j),$$

$$\tag{5.8}$$

where

$$U_k^{i,-}\left(x_1, \varepsilon(j + l_i), x_3, \xi_2 - j\right) = \sum_{m=0}^{k-1} \frac{(\xi_2 - j - l_i)^m}{m!} \frac{\partial^m u_{k-m}^{i,-}}{\partial x_2^m}(x_1, \varepsilon(j + l_i), x_3, \xi_2 - j)$$

$$+ \frac{(\xi_2 - j - l_i)^k}{k!} \frac{\partial^k u_0^{i,-}}{\partial x_2^k}(x_1, \varepsilon(j + l_i), x_3, \xi_2 - j) \tag{5.9}$$

and $\xi_2 = x_2/\varepsilon$ is a "fast" variable.

Again exploiting Taylor's formula, we get

$$k_0(u_\varepsilon(x)) = k_0(u_0^{i,-}(x_1, \varepsilon(j + l_i), x_3)) + \mathcal{O}(\varepsilon), \quad \varepsilon \to 0 \ (x \in \Omega_\varepsilon^{(i)}(j)). \tag{5.10}$$

Substituting (5.8) into (5.1) instead of u_ε, taking into account (2.8), (2.17), (5.7), and the fact that the Laplace operator for the variables $(\tilde{x}, \xi_2)$ reads $\Delta_x = \Delta_{\tilde{x}} + \varepsilon^{-2}\frac{\partial^2}{\partial\xi_2^2}$, and then collecting coefficients at the same powers of ε, we obtain BVPs with respect to ξ_2 for the functions $U_k^{i,-}$.

The function $U_1^{i,-}$ should be a solution to the problem

$$\begin{cases} \partial_{\xi_2\xi_2}^2 U_1^{i,-} = 0, & \xi_2 \in \varepsilon^{-1}I_\varepsilon^{(i)}(j, h_i(r)), \\ \partial_{\xi_2} U_1^{i,-} = 0, & \xi_2 \in \varepsilon^{-1}\partial I_\varepsilon^{(i)}(j, h_i(r)), \end{cases} \tag{5.11}$$

where the variables $\tilde{x}$ are involved here as parameters; the interval $I_\varepsilon^{(i)}(j, h_i(r))$ is defined in Sect. 2.1; $r \in (r_0, r_i)$. Problem (5.11) implies that $U_1^{i,-}$ is independent of ξ_2. Thus $U_1^{i,-}$ is equal to some function $\varphi_1^{(i)}(x_1, \varepsilon(j + l_i), x_3)$, $r \in (r_0, r_i)$, which will be defined later. Then according to (5.9), we have

$$u_1^{i,-}\left(x_1, \varepsilon(j+l_i), x_3, \frac{x_2}{\varepsilon} - j\right) = \varphi_1^{(i)}\left(x_1, \varepsilon(j+l_i), x_3\right)$$
$$- \left(\frac{x_2}{\varepsilon} - j - l_i\right) \partial_{x_2} u_0^{i,-}\left(x_1, \varepsilon(j+l_i), x_3\right).$$
$$(5.12)$$

Boundary-value problems for $U_2^{1,-}$ and $U_2^{2,-}$ look as follows:

$$- \partial_{\xi_2\xi_2}^2 U_2^{1,-} = (\Delta_{\tilde{x}} u_0^{1,-} - k_0(u_0^{1,-}) + f_0)|_{x_2=\varepsilon(j+l_1)}, \qquad \xi_2 \in \varepsilon^{-1} I_\varepsilon^{(1)}(j, h_1(r)),$$
$$\pm \partial_{\xi_2} U_2^{1,-}\big|_{\xi_2=j+l_1\pm\frac{h_1(r)}{2}} = \left(2^{-1}\nabla_{\tilde{x}} h_1 \cdot \nabla_{\tilde{x}} u_0^{1,-} - \delta_{\alpha,1}\kappa_1(u_0^{1,-}) + \delta_{\beta,1} g_0\right)|_{x_2=\varepsilon(j+l_1)},$$

and

$$- \partial_{\xi_2\xi_2}^2 U_2^{2,-} = (\Delta_{\tilde{x}} u_0^{2,-} - \omega(u_0^{2,-}) + f_0)|_{x_2=\varepsilon(j+l_2)}, \qquad \xi_2 \in \varepsilon^{-1} I_\varepsilon^{(2)}(j, h_2(r)),$$
$$\pm \partial_{\xi_2} U_2^{2,-}\big|_{\xi_2=j+l_2\pm\frac{h_2(r)}{2}} = \left(2^{-1}\nabla_{\tilde{x}} h_2 \cdot \nabla_{\tilde{x}} u_0^{2,-} - \kappa_2(u_0^{2,-}) + \delta_{\beta,1} g_0\right)|_{x_2=\varepsilon(j+l_2)}.$$

Here the variables $\tilde{x}$ are parameters. Solvability conditions for these problems read

$$- \operatorname{div}_{\tilde{x}}\left(h_1(r)\nabla_{\tilde{x}} u_0^{1,-}\right) + h_1(r) k_0(u_0^{1,-}) + 2\delta_{\alpha,1}\kappa_1(u_0^{1,-}) = h_1(r) f_0 + 2\delta_{\beta,1} g_0,$$
$$(5.13)$$

where $r \in (r_0, r_1)$, $x_2 = \varepsilon(j+l_1)$, and

$$- \operatorname{div}_{\tilde{x}}\left(h_2(r)\nabla_{\tilde{x}} u_0^{2,-}\right) + h_2(r) k_0(u_0^{2,-}) + 2\kappa_2(u_0^{2,-}) = h_2(r) f_0 + 2\delta_{\beta,1} g_0, \quad (5.14)$$

where $r \in (r_0, r_2)$, $x_2 = \varepsilon(j+l_2)$, respectively.

Substituting (5.8) into the Robin boundary conditions on $\partial\Omega_\varepsilon^{(i)} \cap \{r = r_i\}$ and utilizing (5.10), we obtain the relations

$$\partial_r u_0^{1,-} = 0, \quad r = r_1, \ x_2 = \varepsilon(j+l_1),$$
$$\partial_r u_0^{2,-} + \kappa_2(u_0^{2,-}) = 0, \quad r = r_2, \ x_2 = \varepsilon(j+l_2). \qquad (5.15)$$

In order to find relations on the joint zone Ω', we will use the method of matched asymptotic expansions for the outer expansions (5.5), (5.6), and an inner expansion which is constructed in the next section.

Inner Expansion

If we pass to the "fast" variables $\xi = (\xi_1, \xi_2) = \left(-\dfrac{r - r_0}{\varepsilon}, \dfrac{x_2}{\varepsilon}\right)$ in δ_0-neighborhood of the joint zone Ω' and send ε to zero, then due to the additional assumption made in this chapter the cross section of the periodicity cell is transformed into a domain $\Pi = \Pi^+ \cup \Pi_1^- \cup \Pi_2^-$ (see Fig. 5.1), which is a union of semiinfinite strips

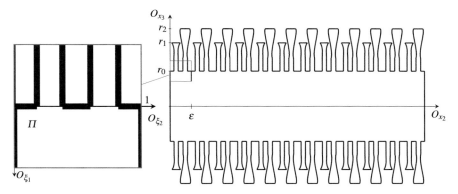

Fig. 5.1 The domain Π and the thick junction Ω_ε

$$\Pi^+ = \{\xi \in \mathbb{R}^2 : \xi_1 > 0, \ \xi_2 \in (0, 1)\},$$
$$\Pi_i^- = \{\xi \in \mathbb{R}^2 : \xi_1 \leq 0, \ \xi_2 \in I_i(r_0)\}, \quad i = 1, 2.$$

The interval $I_i(s)$ is defined in Sect. 2.1.

The Laplace operator for the variables (ξ, θ) reads

$$\Delta = \varepsilon^{-2} \left(\frac{\partial^2}{\partial \xi_1^2} + \frac{\partial^2}{\partial \xi_2^2} \right) - \varepsilon^{-1} \frac{1}{r_0 - \varepsilon \xi_1} \frac{\partial}{\partial \xi_1} + \frac{1}{(r_0 - \varepsilon \xi_1)^2} \frac{\partial^2}{\partial \theta^2}. \tag{5.16}$$

We propose the following ansatz for the solution u_ε in a neighborhood of Ω':

$$u_\varepsilon(x) \approx u_0^+|_{\Omega'}(x) + \varepsilon \Big(- \{\eta(x_2, \theta) Z_1(\xi) + (1 - \eta(x_2, \theta)) Z_2(\xi)\} \partial_r u_0^+|_{\Omega'}(x)$$
$$+ Z_3(\xi) \partial_{x_2} u_0^+|_{\Omega'}(x) \Big) \Big|_{\xi_1 = -\frac{r - r_0}{\varepsilon}, \ \xi_2 = \frac{x_2}{\varepsilon}} + \dots, \tag{5.17}$$

where Z_1, Z_2, Z_3 are 1-periodic with respect to ξ_2 functions in Π, and η is some function. Those functions are unknown and will be defined later. The expansion (5.17) is called an *inner expansion* for the solution u_ε.

Remark 5.1 One of the main difficulties in the construction of asymptotic approximations for solutions of BVPs in thick junctions is to guess an ansatz for the inner expansion. In our case, the summands $\eta Z_1 \partial_r u_0^+|_{\Omega'}$ and $(1 - \eta) Z_2 \partial_r u_0^+|_{\Omega'}$ are to eliminate discrepancies on $\partial \Omega_\varepsilon \cap \{r = r_0\}$. We need two of them due to the fact that we have a thick **two**-level junction. The multipliers η and $1 - \eta$ are to satisfy the matching condition. The summand $Z_3 \partial_{x_2} u_0^+|_{\Omega'}$ is to eliminate discrepancies on $\partial \Omega_\varepsilon^{(i)} \cap \{r_0 < r < r_i\}$ (in a neighborhood of the joint zone).

Substituting (5.17) into the differential equation of problem (5.1) and into the proper boundary conditions, taking into account (5.16), and collecting coefficients of the same powers of ε, we obtain junction-layer problems for the functions $\{Z_i\}$. So, the functions Z_1 and Z_2 are solutions to the problem

$$
\begin{cases}
-\Delta_\xi Z(\xi) = 0, & \xi \in \Pi, \\
\partial_{\xi_2} Z(\xi) = 0, & \xi \in (\partial \Pi_1^- \cup \partial \Pi_2^-) \cap \{\xi \in \mathbb{R}^2 : \xi_1 < 0\}, \\
\partial_{\xi_1} Z(\xi) = 0, & \xi \in \partial \Pi \cap \{\xi \in \mathbb{R}^2 : \xi_1 = 0\}, \\
\partial_{\xi_2}^p Z(\xi_1, 0) = \partial_{\xi_2}^p Z(\xi_1, 1), & p = 0,\, 1,\ \xi_1 > 0,
\end{cases}
\tag{5.18}
$$

and Z_3 is a solution to the problem

$$
\begin{cases}
-\Delta_\xi Z_3(\xi) = 0, & \xi \in \Pi, \\
\partial_{\xi_2} Z_3(\xi) = -1, & \xi \in (\partial \Pi_1^- \cup \partial \Pi_2^-) \cap \{\xi \in \mathbb{R}^2 : \xi_1 < 0\}, \\
\partial_{\xi_1} Z_3(\xi) = 0, & \xi \in \partial \Pi \cap \{\xi \in \mathbb{R}^2 : \xi_1 = 0\}, \\
\partial_{\xi_2}^p Z_3(\xi_1, 0) = \partial_{\xi_2}^p Z(\xi_1, 1), & p = 0,\, 1,\ \xi_1 > 0.
\end{cases}
\tag{5.19}
$$

The main asymptotic relations for solutions to those problems can be obtained from general results on asymptotic behavior of solutions to elliptic BVPs in unbounded domains (see e.g., [63, 66, 122]). However, we can obtain more precise relations for the asymptotics of Z_1, Z_2, Z_3 similarly as was done in [79, 108].

Consider a Sobolev space

$$
H_{\mathrm{loc}}^1(\Pi) = \big\{\varphi : \Pi \mapsto \mathbb{R} : \ \forall R > 0 \ \varphi \in H^1(\Pi_R), \ \varphi(\xi_1, 0) = \varphi(\xi_1, 1) \,\forall\, \xi_1 > 0\big\},
$$

where $\Pi_R = \{\xi \in \Pi : -R < \xi_1 < R\}$.

Statement 1. *There exist two solutions* Z_1, $Z_2 \in H_{\mathrm{loc}}^1(\Pi)$ *to problem* (5.18) *with the differentiable asymptotics*

$$
Z_1(\xi) =
\begin{cases}
\xi_1 + \mathcal{O}(e^{-2\pi \xi_1}), & \xi_1 \to +\infty,\ \xi_2 \in (0,\,1), \\
h_1^{-1}(r_0)\xi_1 + c_1^{(1)} + \mathcal{O}(e^{\pi h_1^{-1}(r_0)\xi_1}), & \xi_1 \to -\infty,\ \xi_2 \in I_1(r_0), \\
c_1^{(2)} + \mathcal{O}(e^{\pi h_2^{-1}(r_0)\xi_1}), & \xi_1 \to -\infty,\ \xi_2 \in I_2(r_0),
\end{cases}
\tag{5.20}
$$

$$
Z_2(\xi) =
\begin{cases}
\xi_1 + \mathcal{O}(e^{-2\pi \xi_1}), & \xi_1 \to +\infty,\ \xi_2 \in (0,\,1), \\
c_2^{(1)} + \mathcal{O}(e^{\pi h_1^{-1}(r_0)\xi_1}), & \xi_1 \to -\infty,\ \xi_2 \in I_1(r_0), \\
h_2^{-1}(r_0)\xi_1 + c_2^{(2)} + \mathcal{O}(e^{\pi h_2^{-1}(r_0)\xi_1}), & \xi_1 \to -\infty,\ \xi_2 \in I_2(r_0),
\end{cases}
\tag{5.21}
$$

where $c_1^{(i)}$, $c_2^{(i)}$, $i = 1,\, 2$, *are some constants.*

Another solution to problem (5.18) *with polynomial growth at infinity can be represented as a linear combination* $c_0 + c_1 Z_1 + c_2 Z_2$.

Statement 2. *There exists a unique solution* $Z_3 \in H_{\mathrm{loc}}^1(\Pi)$ *to problem* (5.19) *with the differentiable asymptotics*

$$Z_3(\xi) = \begin{cases} \mathscr{O}(e^{-2\pi\xi_1}), & \xi_1 \to +\infty, \ \xi_2 \in (0, 1), \\ -\xi_2 + l_1 + c_3^{(1)} + \mathscr{O}(e^{\pi h_1^{-1}(r_0)\xi_1}), & \xi_1 \to -\infty, \ \xi_2 \in I_1(r_0), \\ -\xi_2 + l_2 + c_3^{(2)} + \mathscr{O}(e^{\pi h_2^{-1}(r_0)\xi_1}), & \xi_1 \to -\infty, \ \xi_2 \in I_2(r_0), \end{cases} \quad (5.22)$$

where $c_3^{(1)}$, $c_3^{(2)}$ are some constants.

Another solution to problem (5.19) with polynomial growth at infinity can be represented as $c_0 + Z_3$.

Now let us check the matching conditions for outer expansions (5.5), (5.6), and inner expansion (5.17), namely (see [54, 55]), the main terms of the asymptotics of the outer expansions as $r \to r_0\mp$ have to coincide with the main terms of the asymptotics of the inner expansion as $\xi_1 \to \pm\infty$.

The asymptotics of u_0^+ in a neighborhood of $(r_0, \varepsilon(j + l_i), \theta) \in \Omega'$ as $r \to r_0-$, $x_2 \to \varepsilon(j + l_i)$ $(i = 1, 2)$ are

$$u_0^+ + \varepsilon\big((\xi_2 - j - l_i)\partial_{x_2}u_0^+ - \xi_1\partial_r u_0^+\big), \quad r = r_0, \quad x_2 = \varepsilon(j + l_i).$$

Taking into account the asymptotics of Z_1, Z_2, Z_3 as $\xi_1 \to +\infty$ (see (5.20), (5.21), (5.22)), we see that the matching conditions for expansions (5.5) and (5.17) holds.

The first terms of the asymptotics for the outer expansions (5.6) as $r \to r_0+$, $x_2 \to \varepsilon(j + l_i)$ $(i = 1, 2)$ are

$$u_0^{i,-} + \varepsilon\big(\varphi_1^{(i)} - \xi_1\partial_r u_0^{i,-}\big), \quad r = r_0, \quad x_2 = \varepsilon(j + l_i). \quad (5.23)$$

In accordance with (5.20), (5.21), and (5.22) the first terms of the asymptotics of the inner expansion (5.17) as $\xi_1 \to -\infty$, $\xi_2 \to j + l_1$ are

$$u_0^+ + \varepsilon\Big(- \big\{(h_1^{-1}(r_0)\xi_1 + c_1^{(1)})\eta + c_2^{(1)}(1 - \eta)\big\}\partial_r u_0^+ + c_3^{(1)}\partial_{x_2}u_0^+\Big), \quad (5.24)$$

and as $\xi_1 \to -\infty$, $\xi_2 \to j + l_2$ are

$$u_0^+ + \varepsilon\Big(- \big\{c_1^{(2)}\eta + (h_2^{-1}(r_0)\xi_1 + c_2^{(2)})(1 - \eta)\big\}\partial_r u_0^+) + c_3^{(2)}\partial_{x_2}u_0^+\Big). \quad (5.25)$$

Comparing the first terms in (5.23), (5.24), and (5.25), we obtain the equalities

$$u_0^+ = u_0^{i,-}, \quad r = r_0, \quad x_2 = \varepsilon(j + l_i), \quad i = 1, 2, \quad j = \overline{0, N - 1}; \quad (5.26)$$

the comparing the second ones gives

$$\varphi_1^{(i)} = c_3^{(i)}\partial_{x_2}u_0^+, \quad r = r_0, \quad x_2 = \varepsilon(j + l_i), \ i = 1, 2, \ j = \overline{0, N - 1}, \quad (5.27)$$

and

$$h_1^{-1}(r_0)\eta\partial_r u_0^+ = \partial_r u_0^{1,-}, \quad r = r_0, \quad x_2 = \varepsilon(j + l_1), \quad j = \overline{0,\ N - 1},$$
$$h_2^{-1}(r_0)(1 - \eta)\partial_r u_0^+ = \partial_r u_0^{2,-}, \quad r = r_0, \quad x_2 = \varepsilon(j + l_2), \quad j = \overline{0,\ N - 1}.$$
$$(5.28)$$

Since the points $\{\varepsilon(j + l_i) : j = \overline{0,\ N - 1}\}, i = 1, 2$, form the ε-net in the interval $(0, l)$, we can spread Eqs. (5.12), (5.13), (5.14) into $\Omega^{(i)}, i = 1, 2,$, respectively; relations (5.15) over $\partial\Omega^{(1)} \cap \{r = r_1\}$ and $\partial\Omega^{(2)} \cap \{r = r_2\}$ respectively; and Eqs. (5.26), (5.27), and (5.28) over Ω'. As a result, we deduce

$$\eta(x_2, \theta) = \frac{h_1(r_0)\partial_r u_0^{1,-}}{\partial_r u_0^+}, \quad x \in \Omega' \tag{5.29}$$

(for points where $\partial_r u_0^+|_{\Omega'} = 0$ we set $\eta(x_2, \theta) = 1$);

$$u_0^+ = u_0^{i,-}, \quad \partial_r u_0^+ = h_1(r_0)\partial_r u_0^{1,-} + h_2(r_0)\partial_r u_0^{2,-}, \quad x \in \Omega'; \tag{5.30}$$

and

$$\varphi_1^{(i)}(x) = c_3^{(i)}\partial_{x_2} u_0^{i,-}(x), \quad x \in \Omega^{(i)}, \ i = 1, 2. \tag{5.31}$$

5.1.3 The Homogenized Problem

With the help of the first terms $u_0^+, u_0^{1,-}$, and $u_0^{2,-}$ of the ansatzes (5.5) and (5.6) we define a multivalued function $\mathbf{u}_0 = (u_0^+,\ u_0^{1,-},\ u_0^{2,-})$. Thanks to relations obtained above, we conclude that $\mathbf{u}_0$ should be a solution to the *homogenized problem*

$$\begin{cases} -\Delta_x u_0^+ + k_0(u_0^+) = f_0, & \text{in } \Omega_0, \\ \partial_{x_2}^p u_0^+(x_1, 0, x_3) = \partial_{x_2}^p u_0^+(x_1, l, x_3), & \text{on } \partial\Omega_0 \cap \{r < r_0\}, p = 0, 1, \\ -\operatorname{div}_{\tilde{x}}(h_1\nabla_{\tilde{x}} u_0^{1,-}) + h_1 k_0(u_0^{1,-}) \\ \quad + 2\delta_{\alpha,1}\kappa_1(u_0^{1,-}) = h_1 f_0 + 2\delta_{\beta,1} g_0, & \text{in } \Omega^{(1)}, \\ \partial_\nu u_0^{1,-} = 0, & \text{on } \partial\Omega^{(1)} \cap \{r = r_1\}, \\ -\operatorname{div}_{\tilde{x}}(h_2\nabla_{\tilde{x}} u_0^{2,-}) + h_2 k_0(u_0^{2,-}) \\ \quad + 2\kappa_2(u_0^{2,-}) = h_2 f_0 + 2\delta_{\beta,1} g_0, & \text{in } \Omega^{(2)}, \\ \partial_\nu u_0^{2,-} + \kappa_2(u_0^{2,-}) = 0, & \text{on } \partial\Omega^{(2)} \cap \{r = r_2\}, \\ u_0^+ = u_0^{1,-} = u_0^{2,-}, & \text{on } \Omega', \\ \partial_r u_0^+ = \sum_{i=1}^{2} h_i(r_0)\partial_r u_0^{i,-}, & \text{on } \Omega'. \end{cases} \tag{5.32}$$

Definition 5.2 A function $\mathbf{u} = (u_0, u_1, u_2) \in \widetilde{\mathbf{H}}_{\text{per}}$ is called a weak solution to problem (5.32) if the integral identity

$$\int_{\Omega_0} (\nabla_x u_0 \cdot \nabla_x p_0 + k_0(u_0)p_0)\, dx + \sum_{i=1}^{2} \int_{\Omega^{(i)}} h_i (\nabla_{\tilde{x}} u_i \cdot \nabla_{\tilde{x}} p_i + k_0(u_i)p_i)\, dx$$

$$+ 2\delta_{\alpha,1} \int_{\Omega^{(1)}} \kappa_1(u_1)p_1\, dx + 2 \int_{\Omega^{(2)}} \kappa_2(u_2)p_2\, dx + h_2(r_2) \int_{\partial\Omega^{(2)} \cap \{r=r_2\}} \kappa_2(u_2)p_2\, d\sigma_x$$

$$= \int_{\Omega_0} f_0 p_0\, dx + \sum_{i=1}^{2} \int_{\Omega^{(i)}} (h_i f_0 + 2\delta_{\beta,1} g_0) p_i\, dx$$

holds for all functions $\mathbf{p} = (p_0, p_1, p_2) \in \widetilde{\mathbf{H}}_{\mathrm{per}}$ (the space $\widetilde{\mathbf{H}}_{\mathrm{per}}$ is defined in Sect. 3.4.1).

From the theory of monotone operators (see e.g. [90, 131]) and assumptions (4.5), it follows that there exists a unique weak solution to problem (5.32).

5.1.4 Asymptotic Approximation and Estimates

Consider a smooth cut-off function $\chi_0(r)$, which is equal to 1 if $|r - r_0| < \delta/2$ and to 0 if $|r - r_0| > \delta$, where $\delta \in (0, \delta_0)$ is some fixed number.

With the help of $\mathbf{u}_0$ and the solutions Z_1, Z_2, Z_3 to problems (5.18) and (5.19), respectively, we construct the following approximation function R_ε:

$$R_\varepsilon(x) = R_\varepsilon^+(x) = u_0^+(x) + \varepsilon \chi_0(r) \mathcal{Z}^+(-\frac{r-r_0}{\varepsilon}, \frac{x_2}{\varepsilon}, x_2, \theta), \quad x \in \Omega_0, \quad (5.33)$$

where

$$\mathcal{Z}^+(\xi, x_2, \theta) = -\sum_{i=1}^{2} h_i(r_0) Z_i(\xi) \partial_r u_0^{i,-}|_{\Omega'} + Z_3(\xi)\partial_{x_2} u_0^+|_{\Omega'} + \xi_1 \partial_r u_0^+|_{\Omega'};$$

and

$$R_\varepsilon(x) = R_\varepsilon^{i,-}(x) = u_0^{i,-}(x) + \varepsilon \left(\left(Y_i \left(\frac{x_2}{\varepsilon}\right) + c_3^{(i)} \right) \partial_{x_2} u_0^{i,-}(x) \right.$$

$$\left. + \chi_0(r) \mathcal{Z}^{i,-}\left(-\frac{r-r_0}{\varepsilon}, \frac{x_2}{\varepsilon}, x_2, \theta\right) \right), \quad x \in \Omega_\varepsilon^{(i)}(j),$$

$$(5.34)$$

where

$$\mathscr{L}^{i,-}(\xi, x_2, \theta) = -\sum_{i=1}^{2} h_i(r_0) Z_i(\xi) \partial_r u_0^{i,-}|_{\Omega'}$$

$$+ \left(Z_3(\xi) - Y_i(\xi_2) - c_3^{(i)} \right) \partial_{x_2} u_0^+|_{\Omega'} + \xi_1 \partial_r u_0^{i,-}|_{\Omega'}, \quad i = 1, 2; \quad j = \overline{0, N-1}.$$

In fact we sum the outer expansions with the inner expansion subtracting the common part of their asymptotics, because it is summed twice. Due to (5.29)–(5.31), the approximation function $R_\varepsilon \in H^1(\Omega_\varepsilon)$.

Theorem 5.1 *Suppose that in addition to the assumptions made in Sect. 5.1.1, the following conditions hold:* $f_0 \in H^3(\Omega_0 \cup \Omega^{(2)})$; $\partial_{x_2}^p f_0(x_1, 0, x_3) = \partial_{x_2}^p f_0(x_1, l, x_3)$ *for* $x \in \partial \Omega_0 \cap \{r < r_0\}$ *and* $p = 0, 1$; $g_0 \in H^1(\Omega^{(2)})$ *and it vanishes at a neighborhood of the joint zone* Ω'.

Then for any $\mu > 0$ *there exist positive constants* ε_0, C_1 *such that for all values* $\varepsilon \in (0, \varepsilon_0)$ *the difference between the solution* u_ε *to problem* (5.1) *and the approximating function* R_ε *defined by* (5.33) *and* (5.34) *satisfies the following estimate:*

$$\|u_\varepsilon - R_\varepsilon\|_{H^1(\Omega_\varepsilon)} \leq C_1 \Big(\|f_0 - f_\varepsilon\|_{L^2(\Omega_\varepsilon)} + \delta_{\beta,1} \|g_0 - g_\varepsilon\|_{L^2(\Omega^{(2)})}$$

$$+ \varepsilon^{1-\mu} + \varepsilon^{\alpha-1+\delta_{\alpha,1}} + (1 - \delta_{\beta,1})\varepsilon^{\beta-1} \Big). \quad (5.35)$$

Proof **1.** First we find discrepancies in Ω_0. The first relation in (5.32) implies that

$$\partial_{x_2 x_2}^2 u_0^+(x) = -\Delta_{\tilde{x}} u_0^+(x) + k_0(u_0^+(x)) - f_0(x), \quad x \in \Omega_0,$$

whence due to the second relation in (5.32) we conclude that

$$\partial_{x_2 x_2}^2 u_0^+(x_1, 0, x_3) = \partial_{x_2 x_2}^2 u_0^+(x_1, l, x_3), \quad x \in \partial \Omega_0 \cap \{r < r_0\}.$$

Consequently, according to the properties of the solutions Z_1, Z_2, Z_3, the function R_ε^+ satisfies the boundary conditions of problem (5.1) on $\partial \Omega_\varepsilon \cap \partial \Omega_0$.

It is easy to verify that

$$\Delta_x(\chi_0(r)\varphi(x)) = \text{div}_{\tilde{x}}(\varphi(x)\nabla_{\tilde{x}}\chi_0(r)) + \nabla_{\tilde{x}}\chi_0(r) \cdot \nabla_{\tilde{x}}\varphi(x) + \chi_0(r)\Delta_x \varphi(x). \tag{5.36}$$

Utilizing (5.16), (5.32), and (5.36), we obtain the equality

$$-\Delta_x R_\varepsilon^+(x) + k_0(R_\varepsilon^+) - f_\varepsilon(x) = f_0(x) - f_\varepsilon(x) + k_0(R_\varepsilon^+) - k_0(u_0^+(x))$$

$$+ \chi_0(r)(r^{-1}\partial_{\xi_1} \mathscr{L}^+(\xi, x_2, \theta) - 2\partial_{\xi_2 x_2}^2 \mathscr{L}^+(\xi, x_2, \theta))$$

$$- \varepsilon \, \text{div}_{\tilde{x}}(\mathscr{L}^+|_{\xi_1 = -(r-r_0)/\varepsilon} \nabla_{\tilde{x}} \chi_0(r)) + \chi_0'(r)\partial_{\xi_1} \mathscr{L}^+(\xi, x_2, \theta) \tag{5.37}$$

$$- \varepsilon \chi_0(r)\partial_{x_2 x_2}^2 \mathscr{L}^+(\xi, x_2, \theta) - \varepsilon r^{-2}\chi_0(r)\partial_{\theta\theta}^2 \mathscr{L}^+(\xi, x_2, \theta) \quad x \in \Omega_0,$$

where $\xi = (\xi_1, \xi_2)$, $\xi_1 = -\frac{r-r_0}{\varepsilon}$, $\xi_2 = \frac{x_2}{\varepsilon}$.

Multiplying this equality by a test function $\psi \in H^1_{per}(\Omega_\varepsilon)$ and then integrating it over Ω_0, we get

$$\int_{\Omega_0}(\nabla_x R^+_\varepsilon \cdot \nabla_x \psi + k_0(R^+_\varepsilon)\psi)\, dx - \int_{\Omega^{(1)}_\varepsilon \cap \{r=r_0\} \cup \Omega^{(2)}_\varepsilon \cap \{r=r_0\}} \partial_r R^+_\varepsilon \psi\, d\sigma_x - \int_{\Omega_0} f_\varepsilon \psi\, dx$$
$$= I^+_0(\varepsilon, \psi) + \ldots + I^+_3(\varepsilon, \psi) + I^+_7(\varepsilon, \psi),$$

where

$$I^+_0(\varepsilon, \psi) = \int_{\Omega_0}(f_0 - f_\varepsilon)\psi\, dx,$$
$$I^+_1(\varepsilon, \psi) = \int_{\Omega_0} \chi_0(r^{-1}\partial_{\xi_1}\mathscr{L}^+ - \partial^2_{x_2\xi_2}\mathscr{L}^+)\psi\, dx,$$
$$I^+_2(\varepsilon, \psi) = \varepsilon\int_{\Omega_0}\mathscr{L}^+ \nabla_{\tilde{x}}\chi_0 \cdot \nabla_{\tilde{x}}\psi\, dx + \int_{\Omega_0}\chi'_0\partial_{\xi_1}\mathscr{L}^+\psi\, dx,$$

$$I^+_3(\varepsilon, \psi) = \varepsilon\int_{\Omega_0}\chi_0\partial_{x_2}\mathscr{L}^+\partial_{x_2}\psi\, dx + \varepsilon\int_{\Omega_0}r^{-2}\chi_0\partial_\theta\mathscr{L}^+\partial_\theta\psi\, dx,$$
$$I^+_7(\varepsilon, \psi) = \int_{\Omega_0}(k_0(R^+_\varepsilon) - k_0(u^+_0))\psi\, dx.$$

2. Now let us find discrepancies in the thin discs. Direct calculations show that

$$\partial_r R^{1,-}_\varepsilon = 0 \qquad\qquad\qquad\qquad\qquad \text{on } \partial\Omega^{(1)}_\varepsilon \cap \{r = r_1\},$$
$$\partial_r R^{2,-}_\varepsilon = -\kappa_2(u^{2,-}_0) - \varepsilon\left(Y_2\left(\frac{x_2}{\varepsilon}\right) + c^{(2)}_3\right)\partial_{x_2}\kappa_2(u^{2,-}_0) \text{ on } \partial\Omega^{(2)}_\varepsilon \cap \{r = r_2\},$$

$$(5.38)$$

and

$$\partial_r R^{i,-}_\varepsilon = \varepsilon(Y_i\left(\frac{x_2}{\varepsilon}\right) + z^{(i)}_3)\partial^2_{rx_2}u^{i,-}_0 + \partial_r R^+_\varepsilon, \quad x \in \Omega^{(i)}_\varepsilon \cap \{r = r_0\}, \ i = 1, 2.$$
$$(5.39)$$

Bearing in mind (2.8) and the fact that the functions h_i are constant in the neighborhood of Ω_0, we deduce that

$$\partial_\nu R^{i,-}_\varepsilon = \varepsilon(N^{(i)}_\varepsilon(r))^{-1}\left(\pm\left(Y_i\left(\frac{x_2}{\varepsilon}\right) + c^{(i)}_3\right)\partial^2_{x_2x_2}u^{i,-}_0 \pm \chi_0\partial_{x_2}(\mathscr{L}^{i,-}|_{\xi_2=x_2/\varepsilon})\right.$$
$$\left. - \frac{1}{2}\nabla_{\tilde{x}}h_i \cdot \nabla_{\tilde{x}}\left(u^{i,-}_0 + \varepsilon(Y_i\left(\frac{x_2}{\varepsilon}\right) + c^{(i)}_3)\partial_{x_2}u^{i,-}_0\right)\right) \quad (5.40)$$

for a.e. $x \in \partial\Omega^{(i)}_\varepsilon(j) \cap \{r_0 < r < r_i\}$, $i = 1, 2$, where "+" ("−") in "±" indicates the right (left) part of the lateral surface of jth thin disc.

With the help of (5.16) and (5.32), we get

$$-\Delta_x R^{i,-}_\varepsilon(x) + k_0(R^{i,-}_\varepsilon) - f_\varepsilon(x) = f_0(x) - f_\varepsilon(x) + k_0(R^{i,-}_\varepsilon) - k_0(u^{i,-}_0(x))$$
$$+ \nabla_{\tilde{x}}(\ln h_i(r)) \cdot \nabla_{\tilde{x}}u^{i,-}_0(x) - 2(\delta_{i,1}\delta_{\alpha,1} + \delta_{i,2})h^{-1}_i(r)\kappa_i(u^{i,-}_0(x)) + 2\delta_{\beta,1}h^{-1}_i(r)g_0(x)$$

$$- \varepsilon \operatorname{div}_x \left(\left(Y_i \left(\frac{x_2}{\varepsilon} \right) + z_3^{(i)} \right) \nabla_x (\partial_{x_2} u_0^{i,-}(x)) \right)$$

$$- \varepsilon \Delta_x \left(\chi_0(r) \mathscr{L}^{i,-}(-\frac{r - r_0}{\varepsilon}, \frac{x_2}{\varepsilon}, x_2, \theta) \right), \quad x \in \Omega_\varepsilon^{(i)}, \ i = 1, 2.$$

Using (5.36), we rewrite the last summand similarly as in (5.37). Then multiplying those equalities with a test function $\psi \in H^1_{\mathrm{per}}(\Omega_\varepsilon)$, integrating it over $\Omega_\varepsilon^{(i)}$, and taking into account (2.9), (5.39), (5.40), (5.38), we obtain the identities

$$\int_{\Omega_\varepsilon^{(1)}} (\nabla_x R_\varepsilon^{1,-} \cdot \nabla_x \psi + k_0(R_\varepsilon^{1,-})\psi)\, dx + \varepsilon^\alpha \int_{\partial\Omega_\varepsilon^{(1)} \cap \{r > r_0\}} \kappa_1(R_\varepsilon^{1,-})\psi\, d\sigma_x$$

$$+ \int_{\Omega_\varepsilon^{(1)} \cap \{r = r_0\}} \partial_r R_\varepsilon^+ \psi\, d\sigma_x - \int_{\Omega_\varepsilon^{(1)}} f_\varepsilon \psi\, dx - \varepsilon^\beta \int_{\partial\Omega_\varepsilon^{(1)} \cap \{r > r_0\}} g_\varepsilon \psi\, d\sigma_x$$

$$= I_0^{1,-}(\varepsilon,\,\psi) + \ldots + I_7^{1,-}(\varepsilon,\,\psi)$$

and

$$\int_{\Omega_\varepsilon^{(2)}} (\nabla_x R_\varepsilon^{2,-} \cdot \nabla_x \psi + k_0(R_\varepsilon^{2,-})\psi)\, dx + \varepsilon \int_{\partial\Omega_\varepsilon^{(2)} \cap \{r_0 < r < r_2\}} \kappa_2(R_\varepsilon^{2,-})\psi\, d\sigma_x$$

$$+ \int_{\partial\Omega_\varepsilon^{(2)} \cap \{r = r_2\}} \kappa_2(R_\varepsilon^{2,-})\psi\, d\sigma_x + \int_{\Omega_\varepsilon^{(2)} \cap \{r = r_0\}} \partial_r R_\varepsilon^+ \psi\, d\sigma_x - \int_{\Omega_\varepsilon^{(2)}} f_\varepsilon \psi\, dx$$

$$- \varepsilon^\beta \int_{\partial\Omega_\varepsilon^{(2)} \cap \{r > r_0\}} g_\varepsilon \psi\, d\sigma_x = I_0^{2,-}(\varepsilon,\,\psi) + \ldots + I_7^{2,-}(\varepsilon,\,\psi)$$

for all $\psi \in H^1_{\mathrm{per}}(\Omega_\varepsilon)$, where

$$I_0^{i,-}(\varepsilon,\,\psi) = \int_{\Omega_\varepsilon^{(i)}} (f_0 - f_\varepsilon)\psi\, dx,$$

$$I_1^{i,-}(\varepsilon,\,\psi) = \int_{\Omega_\varepsilon^{(i)}} \chi_0(r^{-1}\partial_{\xi_1}\mathscr{L}^{i,-} - \partial^2_{x_2 \xi_2}\mathscr{L}^{i,-})\psi\, dx,$$

$$I_2^{i,-}(\varepsilon,\,\psi) = \varepsilon \int_{\Omega_\varepsilon^{(i)}} \mathscr{L}^{i,-}\nabla_{\tilde{x}}\chi_0 \cdot \nabla_{\tilde{x}}\psi\, dx + \int_{\Omega_\varepsilon^{(i)}} \chi_0'\partial_{\xi_1}\mathscr{L}^{i,-}\psi\, dx,$$

$$I_3^{i,-}(\varepsilon,\,\psi) = \varepsilon \int_{\Omega_\varepsilon^{(i)}} \chi_0 \partial_{x_2}\mathscr{L}^{i,-}\partial_{x_2}\psi\, dx + \varepsilon \int_{\Omega_\varepsilon^{(i)}} r^{-2}\chi_0 \partial_\theta \mathscr{L}^{i,-}\partial_\theta \psi\, dx,$$

$$I_4^{i,-}(\varepsilon,\,\psi) = \varepsilon \int_{\Omega_\varepsilon^{(i)}} Y_i \left(\frac{x_2}{\varepsilon} \right) \partial_{x_2}(\psi \nabla_{\tilde{x}} u_0^{i,-} \cdot \nabla_{\tilde{x}} \ln h_i)\, dx +$$

$$\varepsilon \int_{\Omega_\varepsilon^{(i)}} \left(Y_i \left(\frac{x_2}{\varepsilon} \right) + z_3^{(i)} \right) \nabla_x (\partial_{x_2} u_0^{i,-}) \cdot \nabla_x \psi\, dx, \quad i = 1, 2,$$

$$I_5^{1,-}(\varepsilon,\,\psi) = 2\delta_{\beta,1} \int_{\Omega_\varepsilon^{(1)}} h_1^{-1} g_0 \psi\, dx - \varepsilon^\beta \int_{\partial\Omega_\varepsilon^{(1)} \cap \{r > r_0\}} g_\varepsilon \psi\, d\sigma_x,$$

$$I_5^{2,-}(\varepsilon,\,\psi) = 2\delta_{\beta,1} \int_{\Omega_\varepsilon^{(2)}} h_2^{-1} g_0 \psi\, dx - \varepsilon^\beta \int_{\partial\Omega_\varepsilon^{(2)} \cap \{r_0 < r < r_2\}} g_\varepsilon \psi\, d\sigma_x,$$

$$I_6^{1,-}(\varepsilon,\,\psi) = -2\delta_{\alpha,1} \int_{\Omega_\varepsilon^{(1)}} h_1^{-1}\kappa_1(u_0^{1,-})\psi\, dx + \varepsilon^\alpha \int_{\partial\Omega_\varepsilon^{(1)} \cap \{r > r_0\}} \kappa_1(R_\varepsilon^{1,-})\psi\, d\sigma_x,$$

$$I_6^{2,-}(\varepsilon, \psi) = -2 \int_{\Omega_\varepsilon^{(2)}} h_2^{-1} \kappa_2(u_0^{2,-}) \psi \, dx + \varepsilon \int_{\partial\Omega_\varepsilon^{(2)} \cap \{r_0 < r < r_2\}} \kappa_2(R_\varepsilon^{2,-}) \psi \, d\sigma_x$$

$$+ \int_{\partial\Omega_\varepsilon^{(2)} \cap \{r=r_2\}} \left(\kappa_2(R_\varepsilon^{2,-}) - \kappa_2(u_0^{2,-}) - \varepsilon \left(Y_2 \left(\frac{x_2}{\varepsilon} \right) + c_3^{(2)} \right) \partial_{x_2} \kappa_2(u_0^{2,-}) \right) \psi \, d\sigma_x,$$

$$I_7^{i,-}(\varepsilon, \psi) = \int_{\Omega_\varepsilon^{(i)}} (k_0(R_\varepsilon^{i,-}) - k_0(u_0^{i,-})) \psi \, dx, \quad i = 1, 2.$$

3. Now we prove the main asymptotic estimate. Summing the integral identities obtained in the previous parts of the proof, we see that the function R_ε satisfies the integral identity

$$\int_{\Omega_\varepsilon} (\nabla_x R_\varepsilon \cdot \nabla_x \psi + k_0(R_\varepsilon) \psi) \, dx + \varepsilon^\alpha \int_{\partial\Omega_\varepsilon^{(1)} \cap \{r>r_0\}} \kappa_1(R_\varepsilon) \psi \, d\sigma_x$$

$$+ \varepsilon \int_{\partial\Omega_\varepsilon^{(2)} \cap \{r_0 < r < r_2\}} \kappa_2(R_\varepsilon) \psi \, d\sigma_x + \int_{\partial\Omega_\varepsilon^{(2)} \cap \{r=r_2\}} \kappa_2(R_\varepsilon) \psi \, d\sigma_x - L_\varepsilon(\psi) = F_\varepsilon(\psi)$$

for every $\psi \in H_{per}^1(\Omega_\varepsilon)$, where $L_\varepsilon(\psi)$ is defined in (5.4), $F_\varepsilon(\psi) = I_0 + \ldots + I_4 + I_5^- + I_6^- + I_7$, $I_m = I_m^+ + I_m^-$, $m \in \{0, \ldots, 4\}$, $I_7 = I_7^+ + I_7^-$, $I_k^- = I_k^{1,-} + I_k^{2,-}$, $k \in \{0, \ldots, 7\}$.

Subtracting (5.3) from the integral identity above, we get

$$\int_{\Omega_\varepsilon} (\nabla_x (R_\varepsilon - u_\varepsilon) \cdot \nabla_x \psi + (k_0(R_\varepsilon) - k_0(u_\varepsilon)) \psi) \, dx$$

$$+ \varepsilon^\alpha \int_{\partial\Omega_\varepsilon^{(1)} \cap \{r>r_0\}} (\kappa_1(R_\varepsilon) - \kappa_1(u_\varepsilon)) \psi \, d\sigma_x + \varepsilon \int_{\partial\Omega_\varepsilon^{(2)} \cap \{r_0 < r < r_2\}} (\kappa_2(R_\varepsilon) - \kappa_2(u_\varepsilon)) \psi \, d\sigma_x$$

$$+ \int_{\partial\Omega_\varepsilon^{(2)} \cap \{r=r_2\}} (\kappa_2(R_\varepsilon) - \kappa_2(u_\varepsilon)) \psi \, d\sigma_x = F_\varepsilon(\psi) \quad \forall \psi \in H_{per}^1(\Omega_\varepsilon). \tag{5.41}$$

Let us estimate $F_\varepsilon(\psi)$. Using Cauchy–Bunyakovsky inequality, we get

$$|I_0(\varepsilon, \psi)| \leq \|f_0 - f_\varepsilon\|_{L^2(\Omega_\varepsilon)} \|\psi\|_{L^2(\Omega_\varepsilon)}.$$

Since $\partial_{\xi_1} \mathscr{L}^+$, $\partial_{x_2\xi_2}^2 \mathscr{L}^+$, $\partial_{\xi_1} \mathscr{L}^{i,-}$, $\partial_{x_2\xi_2}^2 \mathscr{L}^{i,-}$ exponentially decrease as $|\xi_1| \to \infty$ (see (5.20), (5.21), (5.22)), on the grounds of [30, Lemma 3.1] we deduce that

$$\forall \mu > 0 \ \exists C_2 > 0 \ \exists \varepsilon_0 > 0 \ \forall \varepsilon \in (0, \varepsilon_0): \quad |I_1(\varepsilon, \psi)| \leq C_2 \varepsilon^{1-\mu} \|\psi\|_{H^1(\Omega_\varepsilon)}.$$

Integrals in $I_2(\varepsilon, \psi)$ are over the set $\{x \in \Omega_\varepsilon : \delta/2 < |r - r_0| < \delta\}$, where according to (5.20), (5.21), and (5.22) the functions $\mathscr{L}^+$, $\partial_{\xi_1} \mathscr{L}^+$, $\partial_{\xi_1} \mathscr{L}^{i,-}$ are exponentially small. The functions $\mathscr{L}^{i,-}$ can be estimated by some constant c_0. Thus,

$$|I_2(\varepsilon, \psi)| \leq C_3 \varepsilon \|\psi\|_{H^1(\Omega_\varepsilon)}.$$

Integrals in I_3 are over $\{x \in \Omega_\varepsilon : |r - r_0| < \delta\}$ and can be estimated extracting the exponentially decreasing part in the proper integrals and using the Cauchy–Schwartz–Bunyakovsky inequality. Consider, for instance, the integral

$$\left| \int_{\Omega_\varepsilon^{(1)}} \chi_0 \partial_{x_2} \mathscr{Z}^{1,-} \partial_{x_2} \psi \, dx \right| = \left| \int_{\Omega_\varepsilon^{(1)}} \chi_0 \Big\{ \partial_{x_2} \big((\xi_1 - h_1(r_0)(Z_1(\xi) - c_1^{(1)})) \partial_r u_0^{1,-} |_{\Omega'} \right.$$
$$- h_2(r_0) \big(Z_2(\xi) - c_2^{(1)} \big) \partial_r u_0^{2,-} |_{\Omega'} + \big(Z_3(\xi) - Y_1(\xi_2) - c_3^{(1)} \big) \partial_{x_3} u_0^+ |_{\Omega'}$$
$$\left. - c_1^{(1)} h_1(r_0) \partial_r u_0^{1,-} |_{\Omega'} - c_2^{(1)} h_2(r_0) \partial_r u_0^{2,-} |_{\Omega'} \big) \Big\} \Big|_{\xi_1 = -(r-r_0)/\varepsilon, \, \xi_2 = x_2/\varepsilon} \partial_{x_2} \psi \, dx \right|$$
$$\leq c_1 \|\psi\|_{H^1(\Omega_\varepsilon)} \Big(\sqrt{\varepsilon} \big(\|\xi_1 - h_1(r_0)(Z_1 - c_1^{(1)})\|_{L^2(\Pi_1^-)} + \|Z_2 - c_2^{(1)}\|_{L^2(\Pi_1^-)}$$
$$+ \|Z_3 - Y_1 - c_3^{(1)}\|_{L^2(\Pi_1^-)} \big) + \sqrt{|\Omega_\varepsilon^{(1)}|} \Big),$$

where $|\Omega_\varepsilon^{(1)}|$ is the Lebesgue measure of $\Omega_\varepsilon^{(1)}$. Thanks to (5.20), (5.21), and (5.22) we conclude that the norms in the right-hand sides of the last inequality are uniformly bounded with respect to ε. Analogously we can estimate the rest of the summands in $I_3(\varepsilon, \psi)$. As a result, we get the estimate

$$|I_3(\varepsilon, \psi)| \leq C_4 \varepsilon \|\psi\|_{H^1(\Omega_\varepsilon)}.$$

Remark 5.2 The constants C_3 and C_4 depend on

$$\sup_{x \in \Omega'} |\partial_{x_k} u_0^+ |_{\Omega'}|, \quad \sup_{x \in \Omega'} |\partial^2_{x_k x_m} u_0^+ |_{\Omega'}|, \quad \sup_{x \in \Omega'} |\partial_{x_k} u_0^{i,-} |_{\Omega'}|, \quad \sup_{x \in \Omega'} |\partial^2_{x_k x_m} u_0^{i,-} |_{\Omega'}|,$$

where $k, m \in \{1, 2, 3\}$. Extending problem (5.32) periodically with respect to x_2 through the planes $\{x : x_2 = 0\}$ and $\{x : x_2 = l\}$ and taking (4.5) and assumptions for f_0 and g_0 into account, we conclude that these quantities are bounded thanks to results on the smoothness of solutions to semilinear BVPs (see e.g. [132, Sect. 14]).

Theorem's assumptions imply that $\partial_{x_2} u_0^{i,-} \in H^1(\Omega^{(i)})$, $i = 1, 2$. Therefore

$$|I_4^-(\varepsilon, \psi)| \leq c_2 \varepsilon \sum_{i=1}^{2} \big(\|u_0^{i,-}\|_{H^1(\Omega^{(i)})} + \|\partial_{x_2} u_0^{i,-}\|_{H^1(\Omega^{(i)})} \big) \|\psi\|_{H^1(\Omega_\varepsilon)} \leq C_5 \varepsilon \|\psi\|_{H^1(\Omega_\varepsilon)}.$$

With the help of (2.9), (2.17), (2.13), (5.7), (5.42), and (4.9), we deduce

$$|I_5^-(\varepsilon, \psi)| \leq C_6 \begin{cases} (\varepsilon + \|g_0 - g_\varepsilon\|_{L^2(\Omega^{(2)})}) \|\psi\|_{H^1(\Omega_\varepsilon)}, & \beta = 1, \\ \varepsilon^{\beta-1} \|\psi\|_{H^1(\Omega_\varepsilon)}, & \beta > 1; \end{cases}$$

$$|I_6^-(\varepsilon, \psi)| \leq C_7 \begin{cases} \varepsilon \|\psi\|_{H^1(\Omega_\varepsilon)}, & \alpha = 1, \\ (\varepsilon^{\alpha-1} + \varepsilon) \|\psi\|_{H^1(\Omega_\varepsilon)}, & \alpha > 1. \end{cases}$$

It follows from (4.5) that

$$|\kappa(s_1) - \kappa(s_2)| \leq c_3|s_1 - s_2| \quad \forall s_1, s_2 \in \mathbb{R}. \tag{5.42}$$

Using (5.42), we establish that

$$|I_7^{1,-}(\varepsilon, \psi)| \leq c_4\varepsilon \int_{\Omega_\varepsilon^{(1)}} \left| \left(\left(Y_1\left(\frac{x_2}{\varepsilon}\right) + c_3^{(1)} \right) \partial_{x_2} u_0^{1,-} + \chi_0 \mathscr{L}^{1,-} \right) \psi \right| dx$$

$$\leq c_5\varepsilon \left(\|\psi\|_{L^2(\Omega_\varepsilon^{(1)})} + \int_{\Omega_\varepsilon^{(1)}} \chi_0 |\mathscr{L}^{1,-} \psi| \, dx \right).$$

The last integral of the obtained inequality can be estimated by subtracting the exponentially decreasing part and exploiting [30, Lemma 3.1] (similarly as it was done for I_3). Thus, $|I_7^{1,-}(\varepsilon, \psi)| \leq c_6\varepsilon\|\psi\|_{H^1(\Omega_\varepsilon)}$, $\varepsilon \in (0, \varepsilon_0)$. By the same way, we estimate I_7^+ and $I_7^{2,-}$. As a result, we get

$$|I_7(\varepsilon, \psi)| \leq C_8\varepsilon\|\psi\|_{H^1(\Omega_\varepsilon)} \quad \text{for all } \varepsilon \in (0, \varepsilon_0) \text{ and } \psi \in H^1(\Omega_\varepsilon).$$

Thus, for any $\mu > 0$ and for all $\varepsilon \in (0, \varepsilon_0)$

$$|F_\varepsilon(\psi)| \leq C_9\Big(\|f_0 - f_\varepsilon\|_{L^2(\Omega_\varepsilon)} + \delta_{\beta,1}\|g_0 - g_\varepsilon\|_{L^2(\Omega^{(2)})} +$$

$$\varepsilon^{1-\mu} + \varepsilon^{\alpha-1+\delta_{\alpha,1}} + (1 - \delta_{\beta,1})\varepsilon^{\beta-1} + \Big)\|\psi\|_{H^1(\Omega_\varepsilon)}. \tag{5.43}$$

Condition (4.5) provides the inequality:

$$(k_0(s_1) - k_0(s_2))(s_1 - s_2) \geq c_7(s_1 - s_2)^2 \quad \forall s_1, s_2 \in \mathbb{R}. \tag{5.44}$$

Clearly, the same inequality holds for κ_1 and κ_2.

Setting $\psi := R_\varepsilon - u_\varepsilon$ in (5.41) and using (5.43) and (5.44), we deduce (5.35).

Corollary 5.1 *Let assumptions of Theorem 5.1 hold and let* $\mathbf{u} = (u_0^+, u_0^{1,-}, u_0^{2,-})$ *be a weak solution to problem (5.32). Then estimate (5.35) implies that*

$$\|u_\varepsilon - \mathbf{u}|_{\Omega_\varepsilon}\|_{L^2(\Omega_\varepsilon)} \leq C_{10}\Big(\|f_0 - f_\varepsilon\|_{L^2(\Omega_\varepsilon)} + \delta_{\beta,1}\|g_0 - g_\varepsilon\|_{L^2(\Omega^{(2)})}$$

$$+ \varepsilon^{1-\mu} + \varepsilon^{\alpha-1+\delta_{\alpha,1}} + (1 - \delta_{\beta,1})\varepsilon^{\beta-1}\Big).$$

where

$$\mathbf{u}|_{\Omega_\varepsilon}(x) = \begin{cases} u_0^+(x), & x \in \Omega_0, \\ u_0^{1,-}(x), & x \in \Omega_\varepsilon^{(1)}, \\ u_0^{2,-}(x), & x \in \Omega_\varepsilon^{(2)}. \end{cases}$$

5.2   Semilinear Parabolic Problem

5.2.1   Statement of the Problem

For parabolic case, the right-hand sides f_ε and g_ε belong to $L^2((0, T) \times \Omega_\varepsilon)$ and $L^2((0, T) \times \Omega^{(2)})$, respectively. In addition, there exists a weak derivative $\partial_{x_2} g_\varepsilon \in L^2((0, T) \times \Omega^{(2)})$, and

$$\exists\, C_0 > 0 \; \exists\, \varepsilon_0 > 0 \; \forall \varepsilon \in (0, \varepsilon_0) : \; \|g_\varepsilon\|_{L^2((0, T)\times \Omega^{(2)})} + \|\partial_{x_2} g_\varepsilon\|_{L^2((0, T)\times \Omega^{(2)})} < C_0.$$

The functions k_0, κ_1, and κ_2 satisfy conditions (4.5), and the parameters α, $\beta \geq 1$.

Consider the following semilinear parabolic problem with Robin boundary conditions on the surfaces of the thin discs from both levels:

$$
\begin{cases}
\partial_t u_\varepsilon(t, x) - \Delta_x u_\varepsilon + k_0(u_\varepsilon) = f_\varepsilon, & (t, x) \in (0, T) \times \Omega_\varepsilon, \\
\partial_\nu u_\varepsilon(t, x) + \varepsilon^\alpha \kappa_1(u_\varepsilon) = \varepsilon^\beta g_\varepsilon, & (t, x) \in (0, T) \times \partial\Omega_\varepsilon^{(1)} \cap \{r > r_0\}, \\
\partial_\nu u_\varepsilon(t, x) + \varepsilon \kappa_2(u_\varepsilon) = \varepsilon^\beta g_\varepsilon, & (t, x) \in (0, T) \times \partial\Omega_\varepsilon^{(2)} \cap \{r_0 < r < r_2\}, \\
\partial_\nu u_\varepsilon(t, x) + \kappa_2(u_\varepsilon(t, x)) = 0, & (t, x) \in (0, T) \times \partial\Omega_\varepsilon^{(2)} \cap \{r = r_2\}, \\
\partial_\nu u_\varepsilon(t, x) = 0, & (t, x) \in (0, T) \times \partial\Omega_\varepsilon \cap \{r = r_0\}, \\
\partial_{x_2}^p u_\varepsilon(t, x_1, 0, x_3) = \partial_{x_2}^p u_\varepsilon(t, x_1, l, x_3), & (t, x) \in (0, T) \times \partial\Omega_0 \cap \{r < r_0\},\ p = 0, 1, \\
u(0, x) = 0, & x \in \Omega_\varepsilon.
\end{cases}
\tag{5.45}
$$

We introduce the space $W_{T,\mathrm{per}}(\Omega_\varepsilon)$ defined by

$$W_{T,\mathrm{per}}(\Omega_\varepsilon) := \left\{\varphi \in L^2(0, T; H^1_{\mathrm{per}}(\Omega_\varepsilon)) : \; \partial_t \varphi := \varphi' \in L^2(0, T; (H^1_{\mathrm{per}}(\Omega_\varepsilon))^*)\right\}.$$

It is known (see e.g. [42, Sect. 1, Chap. 4]) that $W_{T,\mathrm{per}}(\Omega_\varepsilon) \subset C([0, T]; L^2(\Omega_\varepsilon))$.

Definition 5.3 A function $u_\varepsilon \in W_{T,\mathrm{per}}(\Omega_\varepsilon)$ is called a weak solution to problem (5.45) if for any $\varphi \in H^1_{\mathrm{per}}(\Omega_\varepsilon)$ and a.e. $t \in (0, T)$ the following identity holds:

$$
\langle \partial_t u_\varepsilon, \varphi \rangle_{H^1_{\mathrm{per}}(\Omega_\varepsilon)} + \int_{\Omega_\varepsilon} (\nabla_x u_\varepsilon \cdot \nabla_x \varphi + \omega(u_\varepsilon)\varphi)\, dx + \varepsilon^\alpha \int_{\partial\Omega_\varepsilon^{(1)} \cap \{r > r_0\}} \kappa_1(u_\varepsilon)\varphi\, d\sigma_x
$$

$$
+ \varepsilon \int_{\partial\Omega_\varepsilon^{(2)} \cap \{r_0 < r < r_2\}} \kappa_2(u_\varepsilon)\varphi\, d\sigma_x + \int_{\partial\Omega_\varepsilon^{(2)} \cap \{r = r_2\}} \kappa_2(u_\varepsilon)\varphi\, d\sigma_x = L_\varepsilon(\varphi),
\tag{5.46}
$$

where the functional L_ε is defined in (5.4).

Similarly as in [92, 131], we can show that for every fixed $\varepsilon > 0$ there exists a unique weak solution of problem (5.45).

5.2.2 Formal Asymptotic Expansions

Here the approach of Sect. 5.1.2 is used, but now we have to account for the time variable t. The outer asymptotic expansion for the solution u_ε in the domain Ω_0 has the form (5.5) and in $\Omega_\varepsilon^{(i)}(j)$ has the form (5.6), where all terms additionally depend on t.

Substituting the expansion (5.5) into the first equation of problem (5.45), into the boundary conditions on $\partial\Omega_0 \cap \{r < r_0\}$, and into the initial conditions, using (5.7), and collecting coefficients at the same powers of ε, we derive the relations for the function u_0^+:

$$
\begin{cases}
\partial_t u_0^+(t,x) - \Delta_x u_0^+(t,x) + k_0(u_0^+(t,x)) = f_0(t,x), & (t,x) \in (0,T) \times \Omega_0, \\
\partial_{x_2}^p u_0^+(t,x_1,0,x_3) = \partial_{x_2}^p u_0^+(t,x_1,l,x_3), & (t,x) \in \partial\Omega_0 \cap \{r < r_0\},\ p = 0,1, \\
u_0^+(0,x) = 0, & x \in \Omega_0.
\end{cases}
$$

We rewrite expansion (5.6) in the form (5.8) and substitute it in problem (5.45) instead of u_ε. Bearing in mind (2.8), (5.10), and collecting coefficients at the same powers of ε, we arrive at problems with respect to the variable ξ_2 for $U_k^{i,-}$.

The functions $U_1^{i,-}$, $i = 1, 2$, are solutions to problems (5.11), whence we deduce that they are independent of ξ_2. Thus, $U_1^{i,-}$ is equal to some function $\varphi_1^{(i)}(t, x_1, \varepsilon(j + l_i), x_3)$, which will be defined later. According to (5.9), equality (5.12) takes place.

Problems for $U_2^{1,-}$ and $U_2^{2,-}$ look as follows:

$$
\partial_{\xi_2 \xi_2}^2 U_2^{1,-} = \big(\partial_t u_0^{1,-} - \Delta_{\tilde{x}} u_0^{1,-} + k_0(u_0^{1,-}) - f_0\big)|_{x_2 = \varepsilon(j+l_1)}, \quad \xi_2 \in \varepsilon^{-1} I_\varepsilon^{(1)}(j, h_1(r)),
$$
$$
\pm \partial_{\xi_2} U_2^{1,-}\big|_{\xi_2 = j+l_1 \pm \frac{h_1(r)}{2}} = \big(2^{-1}\nabla_{\tilde{x}} h_1 \cdot \nabla_{\tilde{x}} u_0^{1,-} - \delta_{\alpha,1}\kappa_1(u_0^{1,-}) + \delta_{\beta,1} g_0\big)|_{x_2 = \varepsilon(j+l_1)},
$$

and

$$
\partial_{\xi_2 \xi_2}^2 U_2^{2,-} = \big(\partial_t u_0^{2,-} - \Delta_{\tilde{x}} u_0^{2,-} + k_0(u_0^{2,-}) - f_0\big)|_{x_2 = \varepsilon(j+l_2)}, \quad \xi_2 \in \varepsilon^{-1} I_\varepsilon^{(2)}(j, h_2(r)),
$$
$$
\pm \partial_{\xi_2} U_2^{2,-}\big|_{\xi_2 = j+l_2 \pm \frac{h_2(r)}{2}} = \big(2^{-1}\nabla_{\tilde{x}} h_2 \cdot \nabla_{\tilde{x}} u_0^{2,-} - \kappa_2(u_0^{2,-}) + \delta_{\beta,1} g_0\big)|_{x_2 = \varepsilon(j+l_2)},
$$

respectively. Solvability conditions for these problems read

$$
h_1(r)\partial_t u_0^{1,-} - \operatorname{div}_{\tilde{x}}(h_1(r)\nabla_{\tilde{x}} u_0^{1,-}) + h_1(r)k_0(u_0^{1,-}) + 2\delta_{\alpha,1}\kappa_1(u_0^{1,-}) = h_1(r)f_0 + 2\delta_{\beta,1} g_0
$$

where $t \in (0,T)$, $r \in (r_0, r_1)$, $x_2 = \varepsilon(j + l_1)$, and

$$
h_2(r)\partial_t u_0^{2,-} - \operatorname{div}_{\tilde{x}}(h_2(r)\nabla_{\tilde{x}} u_0^{2,-}) + h_2(r)k_0(u_0^{2,-}) + 2\kappa_2(u_0^{2,-}) = h_2(r)f_0 + 2\delta_{\beta,1} g_0
$$

where $t \in (0,T)$, $r \in (r_0, r_2)$, $x_2 = \varepsilon(j + l_2)$, respectively.

Substituting (5.8) into the Robinr boundary conditions on $\partial\Omega_\varepsilon^{(i)} \cap \{r = r_i\}$ we get (5.15). Substituting (5.8) into the initial conditions of problem (5.45) we obtain

$$u_0^{i,-}(0, x) = 0, \quad r \in (r_0, r_i), \quad x_2 = \varepsilon(j + l_i), \quad i = 1, 2.$$

In order to find relations on the joint zone Ω', we use the method of matched asymptotic expansions.

We seek for the leading terms of the inner expansion for the u_ε at a neighborhood of the joint zone Ω' in the form

$$u_\varepsilon(t, x) \approx u_0^+|_{\Omega'}(t, x) + \varepsilon\Big(Z_3(\xi)\partial_{x_2}u_0^+|_{\Omega'}(t, x)$$
$$- \big\{\eta(t, x_2, \theta)Z_2(\xi) + (1 - \eta(t, x_2, \theta))Z_1(\xi)\big\}\partial_r u_0^+|_{\Omega'}(t, x)\Big)\Big|_{\xi_1 = -\frac{r-r_0}{\varepsilon},\, \xi_2 = \frac{x_2}{\varepsilon}} + \cdots,$$

$$(5.47)$$

where Z_1, Z_2, Z_3 are solutions to problem (5.18) and (5.19) with the asymptotics (5.20), (5.21), and (5.22), respectively.

Matching the outer expansions with the inner one and repeating assertions of Sect. 5.1.2, we derive the transmission conditions (5.30) and define the functions η and $\varphi_1^{(i)}$, $i = 1, 2$, (see (5.29)) and (5.31)).

5.2.3 The Homogenized Problem

Obtained relations for the leading terms u_0^+, $u_0^{1,-}$, and $u_0^{2,-}$ constitute the *homogenized problem* for problem (5.45):

$$\begin{cases}
\partial_t u_0^+ - \Delta_x u_0^+ + k_0(u_0^+) = f_0, & (t, x) \in (0, T) \times \Omega_0, \\
\partial_{x_2}^p u_0^+(t, x_1, 0, x_3) = \partial_{x_2}^p u_0^+(t, x_1, l, x_3), \ p = 0, 1, & (t, x) \in (0, T) \times \partial\Omega_0 \cap \{r < r_0\}, \\
h_1\partial_t u_0^{1,-} - \mathrm{div}_{\bar{x}}(h_1\nabla_{\bar{x}}u_0^{1,-}) + h_1 k_0(u_0^{1,-}) \\
\quad + 2\delta_{\alpha,1}\kappa_1(u_0^{1,-}) = h_1 f_0 + 2\delta_{\beta,1}g_0, & (t, x) \in (0, T) \times \Omega^{(1)}, \\
\partial_\nu u_0^{1,-} = 0, & (t, x) \in (0, T) \times \partial\Omega^{(1)} \cap \{r = r_1\}, \\
h_2\partial_t u_0^{2,-} - \mathrm{div}_{\bar{x}}(h_2\nabla_{\bar{x}}u_0^{2,-}) + h_2 k_0(u_0^{2,-}) \\
\quad + 2\kappa_2(u_0^{2,-}) = h_2 f_0 + 2\delta_{\beta,1}g_0, & (t, x) \in (0, T) \times \Omega^{(2)}, \\
\partial_\nu u_0^{2,-} + \kappa_2(u_0^{2,-}) = 0, & (t, x) \in (0, T) \times \partial\Omega^{(2)} \cap \{r = r_2\}, \\
u_0^+ = u_0^{1,-} = u_0^{2,-}, & (t, x) \in (0, T) \times \Omega', \\
\partial_r u_0^+ = \sum_{i=1}^{2} h_i(r_0)\partial_r u_0^{i,-}, & (t, x) \in (0, T) \times \Omega', \\
\mathbf{u}_0(0, x) = \mathbf{0}.
\end{cases}$$

$$(5.48)$$

where $\mathbf{u}_0 = (u_0^+, u_0^{1,-}, u_0^{2,-})$. Consider a space of multivalued functions

$$\mathbf{W}_{T,\mathrm{per}} := \Big\{\mathbf{p} = (p_0, p_1, p_2) \in L^2(0, T; \widetilde{\mathbf{H}}_{\mathrm{per}}) : \partial_t\mathbf{p} := \mathbf{p}' \in L^2(0, T; \widetilde{\mathbf{H}}_{\mathrm{per}}^*)\Big\}.$$

Definition 5.4 A function $\mathbf{u} = (u_0, u_1, u_2) \in L^2(0, T; \widetilde{\mathbf{H}}_{\mathrm{per}})$ is called a weak solution to problem (5.48) if the following integral identity holds for every $\mathbf{p} \in \mathbf{W}_{T,\mathrm{per}}$:

$$\int_{\Omega_0} u_0(T, x) p_0(T, x) \, dx + \sum_{i=1}^{2} \int_{\Omega^{(i)}} h_i(r) u_i(T, x) p_i(T, x) \, dx$$

$$+ \int_0^T \left(-\langle u_0, \partial_t p_0 \rangle_{H^1_{per}(r_0)} - \sum_{i=1}^{2} \langle h_i u_i, \partial_t p_i \rangle_{\widetilde{H}^1(\Omega^{(i)})} + \int_{\Omega_0} (\nabla_x u_0 \cdot \nabla_x p_0 + k_0(u_0) p_0) \, dx \right.$$

$$+ \sum_{i=1}^{2} \int_{\Omega^{(i)}} h_i(r) (\nabla_{\tilde{x}} u_i \cdot \nabla_{\tilde{x}} p_i + k_0(u_i) p_i) \, dx + h_2(r_0) \int_{\partial \Omega^{(2)} \cap \{r=r_2\}} \kappa_2(u_2) p_2 \, d\sigma_x$$

$$\left. + 2\delta_{\alpha,1} \int_{\Omega^{(1)}} \kappa_1(u_1) p_1 \, dx + 2 \int_{\Omega^{(2)}} \kappa_2(u_2) p_2 \, dx \right) dt = \int_0^T \mathsf{L}(\mathbf{p}) \, dt,$$

where the space $\widetilde{\mathbf{H}}_{per}$ is defined in Sect. 3.4.1 and the functional

$$\mathsf{L}(\mathbf{p}) = \int_{\Omega_0} f_0 p_0 \, dx + \sum_{i=1}^{2} \int_{\Omega^{(i)}} (h_i(r) f_0 + 2\delta_{\beta,1} g_0) p_i \, dx.$$

Using the theory of monotone operators (see e.g. [92, 131]), we can state that there exists a unique weak solution to problem (5.48).

5.2.4 Asymptotic Approximation and Estimates

With the help of the solution $\mathbf{u}_0$ to the homogenized problem (5.48), the solutions Z_1, Z_2, Z_3 to problems (5.18) and (5.19), respectively, and the cut-off function χ_0 defined in Sect. 5.1.4, we construct the approximation function $R_\varepsilon(t, x)$ by formulas (5.33) and (5.34) (the dependence on the variable t only in terms containing $u_0^+(t, x)$, $u_0^{1,-}(t, x)$, $u_0^{2,-}(t, x)$ and their derivatives).

Theorem 5.2 *Suppose that in addition to the assumptions made in Sect. 5.2.1, the following conditions hold:* $f_0 \in C^3([0, T] \times \overline{\Omega_0 \cup \Omega^{(2)}})$; $g_0 \in C^3([0, T] \times \overline{\Omega^{(2)}})$ *and it vanishes at a neighborhood of the joint zone* Ω'; $\partial_{x_2}^p f_0(t, x_1, 0, x_3) = \partial_{x_2}^p f_0(t, x_1, l, x_3)$ *for* $(t, x) \in [0, T] \times \partial \Omega_0 \cap \{r < r_0\}$ *and* $p = 0, 1$; $f_0(0, x) = g_0(0, x) = 0$. *Then for any* $\mu > 0$ *there exist positive constants* ε_0, C_1 *such that for all values* $\varepsilon \in (0, \varepsilon_0)$

$$\|u_\varepsilon - R_\varepsilon\|_{L^2(0,T;H^1(\Omega_\varepsilon))} + \max_{t \in [0, T]} \|u_\varepsilon(t, \cdot) - R_\varepsilon(t, \cdot)\|_{L^2(\Omega_\varepsilon)} \leq C_1 \left(\varepsilon^{1-\mu} + \varepsilon^{\alpha-1+\delta_{\alpha,1}} \right.$$

$$\left. + (1 - \delta_{\beta,1}) \varepsilon^{\beta-1} + \|f_0 - f_\varepsilon\|_{L^2((0,T) \times \Omega_\varepsilon)} + \delta_{\beta,1} \|g_0 - g_\varepsilon\|_{L^2((0,T) \times \Omega^{(2)})} \right). \tag{5.49}$$

Proof **1**. Similar as in the first item of the proof of Theorem 5.1 we verify that R_ε^+ satisfies all of the boundary conditions of problem (5.45) on $\partial \Omega_\varepsilon \cap \partial \Omega_0$.

Using (5.16), (5.36), and (5.48), we derive that

$$
\partial_t R_\varepsilon^+(t, x) - \Delta_x R_\varepsilon^+(t, x) - f_\varepsilon(t, x) = f_0(t, x) - f_\varepsilon(t, x) - k_0(u_0^+(t, x))
$$
$$
+ \varepsilon \chi_0(r) \partial_t \mathscr{Z}^+(t, \xi, x_2, \theta)\big|_{\xi_1 = -\frac{r-r_0}{\varepsilon}, \xi_2 = \frac{x_2}{\varepsilon}} - \varepsilon \Delta_x \big(\chi_0(r) \mathscr{Z}^+(t, -\frac{r - r_0}{\varepsilon}, \frac{x_2}{\varepsilon}, x_2, \theta)\big)
$$

in $(0, T) \times \Omega_0$. Then we rewrite the last summand similarly as in Theorem 5.1, multiply the last identity by a test function $\psi \in W_{T,\mathrm{per}}(\Omega_\varepsilon)$, and integrate it over $(0, \tau) \times \Omega_0$, where $\tau \in (0, T)$ is an arbitrary number. As a result, we get

$$
\int_{\Omega_0} R_\varepsilon^+(\tau, x)\psi(\tau, x)\, dx + \int_0^\tau \Big(-\langle R_\varepsilon^+, \partial_t \psi\rangle_{H^1_{\mathrm{per}}(r_0)} + \int_{\Omega_0} (\nabla_x R_\varepsilon^+ \cdot \nabla_x \psi + k_0(R_\varepsilon^+)\psi)\, dx
$$

$$
- \int_{\Omega_\varepsilon^{(1)} \cap \{r=r_0\} \cup \Omega_\varepsilon^{(2)} \cap \{r=r_0\}} \partial_r R_\varepsilon^+ \psi\, d\sigma_x - \int_{\Omega_0} f_\varepsilon \psi\, dx \Big) dt
$$

$$
= \int_0^\tau \big(I_0^+(\varepsilon, \psi, t) + \ldots + I_4^+(\varepsilon, \psi, t) + I_7^+(\varepsilon, \psi, t) + I_8^+(\varepsilon, \psi, t) \big)\, dt.
$$

Here $I_0^+, \ldots, I_4^+$, and I_7^+ are defined in item **1** of the proof of Theorem 5.1 and

$$
I_8^+(\varepsilon, \psi, t) = \varepsilon \int_{\Omega_0} \chi_0 \partial_t \mathscr{Z}^+ \psi\, dx.
$$

2. Now we find discrepancies in the thin discs. Direct calculations show that for a.e. $t \in (0, T)$ equalities (5.38), (5.39), and (5.40) hold. Using relations (5.16), (5.48), and (5.36), we derive that

$$
\partial_t R_\varepsilon^{i,-}(t, x) - \Delta_x R_\varepsilon^{i,-}(t, x) - f_\varepsilon(t, x) = f_0(t, x) - f_\varepsilon(t, x) - k_0(u_0^{i,-}(t, x))
$$
$$
+ \nabla_{\bar{x}}(\ln h_i(r)) \cdot \nabla_{\bar{x}} u_0^{i,-}(t, x) - 2(\delta_{i,1}\delta_{\alpha,1} + \delta_{i,2})h_i^{-1}(r)\kappa_i(u_0^{i,-}(t, x)) + 2\delta_{\beta,1}h_i^{-1}(r)g_0(t, x)
$$
$$
- \varepsilon \operatorname{div}_x \Big(\big(Y_i\big(\frac{x_2}{\varepsilon}\big) + c_3^{(i)}\big)\nabla_x(\partial_{x_2} u_0^{i,-}(t, x))\Big)
$$
$$
+ \varepsilon \Big(\big(Y_i\big(\frac{x_2}{\varepsilon}\big) + c_3^{(i)}\big)\partial_{tx_2}^2 u_0^{i,-} + \chi_0(r)\partial_t \mathscr{Z}^{i,-}(t, \xi, x_2, \theta)\big|_{\xi_1 = -\frac{r-r_0}{\varepsilon}, \xi_2 = \frac{x_2}{\varepsilon}}\Big)
$$
$$
- \varepsilon \Delta_x\big(\chi_0(r)\mathscr{Z}^{i,-}(t, -\frac{r - r_0}{\varepsilon}, \frac{x_2}{\varepsilon}, x_2, \theta)\big), \quad (t, x) \in (0, T) \times \Omega_\varepsilon^{(i)}, \ i = 1, 2.
$$

We rewrite the last summand similarly as in Theorem 5.1, then multiply the resulting equality by a test function $\psi \in W_{T,\mathrm{per}}(\Omega_\varepsilon)$, and integrate it over $(0, \tau) \times \Omega_\varepsilon^{(i)}$. As a result, by the same way as in **2** and **3** items of the proof of Theorem 5.1, we get

$$
\int_{\Omega_\varepsilon} R_\varepsilon(\tau, x)\psi(\tau, x)\, dx + \int_0^\tau \Big(-\langle R_\varepsilon, \partial_t \psi\rangle_{H^1_{\mathrm{per}}(\Omega_\varepsilon)} + \int_{\Omega_\varepsilon} (\nabla_x R_\varepsilon \cdot \nabla_x \psi + k_0(R_\varepsilon)\psi)\, dx
$$

$$
+ \varepsilon^\alpha \int_{\partial\Omega_\varepsilon^{(1)} \cap \{r > r_0\}} \kappa_1(R_\varepsilon)\psi\, d\sigma_x + \varepsilon \int_{\partial\Omega_\varepsilon^{(2)} \cap \{r_0 < r < r_2\}} \kappa_2(R_\varepsilon)\psi\, d\sigma_x
$$

$$+ \int_{\partial \Omega_\varepsilon^{(2)} \cap \{r = r_2\}} \kappa_2(R_\varepsilon) \psi \, d\sigma_x - \int_{\Omega_\varepsilon} f_\varepsilon \psi \, dx$$

$$- \varepsilon^\beta \int_{\partial \Omega_\varepsilon^{(1)} \cap \{r > r_0\} \cup \partial \Omega_\varepsilon^{(2)} \cap \{r_0 < r < r_2\}} g_\varepsilon \psi \, d\sigma_x \Bigg) dt = \int_0^\tau \sum_{m=0}^{8} I_i(\varepsilon, \, \psi, \, t) \, dt$$

for every $\psi \in W_{T,\mathrm{per}}(\Omega_\varepsilon)$ and any $\tau \in (0, \, T)$. Here $I_m = I_m^+ + I_m^-$, $m \in \{0, \dots, 4\} \cup \{7, 8\}$, $I_k^- = I_k^{1,-} + I_k^{2,-}$, $k \in \{0, \dots, 7\}$, are defined similarly as in the proof of Theorem 5.1, and $I_8^- = I_8^{1,-} + I_8^{2,-}$,

$$I_8^{i,-}(\varepsilon, \, \psi, \, t) = \varepsilon \int_{\Omega_\varepsilon^{(i)}} \left(\left(Y_i \left(\frac{x_2}{\varepsilon} \right) + z_3^{(i)} \right) \partial_{x_2 t}^2 u_0^{i,-} + \chi_0 \partial_t \mathscr{Z}^{i,-} \right) \psi \, dx, \quad i = 1, 2.$$

3. Subtracting (5.46) from the integral identity above, we obtain

$$\int_{\Omega_\varepsilon} (R_\varepsilon(\tau, \, x) - u_\varepsilon(\tau, \, x)) \psi(\tau, \, x) \, dx + \int_0^\tau \Bigg(- \langle R_\varepsilon - u_\varepsilon, \, \partial_t \psi \rangle_{H_{\mathrm{per}}^1(\Omega_\varepsilon)}$$

$$+ \int_{\Omega_\varepsilon} \left(\nabla_x(R_\varepsilon - u_\varepsilon) \cdot \nabla_x \psi + (k_0(R_\varepsilon) - k_0(u_\varepsilon)) \psi \right) dx$$

$$+ \varepsilon \int_{\partial \Omega_\varepsilon^{(2)} \cap \{r_0 < r < r_2\}} (\kappa_2(R_\varepsilon) - \kappa_2(u_\varepsilon)) \psi \, d\sigma_x + \int_{\partial \Omega_\varepsilon^{(2)} \cap \{r = r_2\}} (\kappa_2(R_\varepsilon) - \kappa_2(u_\varepsilon)) \psi \, d\sigma_x$$

$$+ \varepsilon^\alpha \int_{\partial \Omega_\varepsilon^{(1)} \cap \{r > r_0\}} (\kappa_1(R_\varepsilon) - \kappa_1(u_\varepsilon)) \psi \, d\sigma_x \Bigg) dt = P_\varepsilon(\psi, \, \tau)$$

$$(5.50)$$

for all $\psi \in W_{T,\mathrm{per}}(\Omega_\varepsilon)$ and $\tau \in (0, \, T)$. Here, $P_\varepsilon(\psi, \, \tau) = \int_0^\tau \sum_{m=0}^{8} I_i(\varepsilon, \, \psi, \, t)) \, dt$. In fact it remains to estimate I_8. Subtracting the exponentially decreasing part, similarly as it was done for I_3 in the proof of Theorem 5.1, we deduce

$$|I_8(\varepsilon, \, \psi, \, \tau)| \le C_2 \varepsilon \|\psi\|_{L^2((0,\tau) \times \Omega_\varepsilon)}.$$

Remark 5.3 The constant C_2 depends on

$$\sup_{(t, \, x) \in (0, \, T) \times \Omega'} \left| \partial_{tx_j}^2 u_0^+ |_{\Omega'} \right|, \qquad \sup_{(t, \, x) \in (0, \, T) \times \Omega'} \left| \partial_{tx_j}^2 u_0^{i,-} |_{\Omega'} \right|, \qquad i = 1, 2, \ j = 1, 2, 3.$$

These quantities are bounded, which is explained similarly as in the Remark 5.2 but now we should cite [132, Sect. 15] or [68, Chap. 5, Sects. 6 and 7].

Taking into account estimates obtained for $\{I_m\}$ in the proof of Theorem 5.1, we have that for any $\mu > 0$ there exists $\varepsilon_0 > 0$ such that for any $\tau \in (0, \, T)$ and $\varepsilon \in (0, \, \varepsilon_0)$

$$|P_\varepsilon(\psi, \tau)| \le C_3 \Big(\|f_\varepsilon - f_0\|_{L^2((0,T)\times\Omega_\varepsilon)} + \varepsilon^{1-\mu} + \varepsilon^{\alpha-1+\delta_{\alpha,1}}$$

$$+ (1 - \delta_{\beta,1})\varepsilon^{\beta-1} + \delta_{\beta,1}\|g_\varepsilon - g_0\|_{L^2((0,T)\times\Omega^{(2)})} \Big) \|\psi\|_{L^2(0,T;H^1(\Omega_\varepsilon))} \quad (5.51)$$

Setting $\psi := R_\varepsilon - u_\varepsilon$ in (5.50), with the help of (5.44) we obtain

$$\frac{1}{2}\|u_\varepsilon(\tau, \cdot) - R_\varepsilon(\tau, \cdot)\|^2_{L^2(\Omega_\varepsilon)} + \|u_\varepsilon - R_\varepsilon\|^2_{L^2(0,\tau;H^1(\Omega_\varepsilon))} \le c_0 |P_\varepsilon(R_\varepsilon - u_\varepsilon, \tau)|$$

for all $\tau \in (0, T)$. From this inequality, thanks to (5.51) it follows (5.49).

Corollary 5.2 *Let assumptions of Theorem 5.2 hold and let* $\mathbf{u} = (u_0^+, u_0^{1,-}, u_0^{2,-})$ *be a weak solution to problem* (5.48). *Then estimate* (5.49) *implies that*

$$\max_{t\in[0,T]} \|u_\varepsilon(t, \cdot) - \mathbf{u}|_{\Omega_\varepsilon}(t, \cdot)\|_{L^2(\Omega_\varepsilon)} \le C_4 \Big(\|f_0 - f_\varepsilon\|_{L^2((0,T)\times\Omega_\varepsilon)} + \varepsilon^{1-\mu} + \varepsilon^{\alpha-1+\delta_{\alpha,1}}$$

$$+ (1 - \delta_{\beta,1})\varepsilon^{\beta-1} + \delta_{\beta,1}\|g_0 - g_\varepsilon\|_{L^2((0,T)\times\Omega^{(2)})} \Big)$$

where $\mathbf{u}|_{\Omega_\varepsilon}(t, x)$ *is defined similarly as in Corollary* 5.1.

5.3 Conclusions to this Chapter

In contrast to the results of Chaps. 2, 3, and 4, where only the convergence theorems were proved, here the approximation functions are constructed and asymptotic estimates in the proper Sobolev spaces are proved for the solutions both to elliptic and parabolic semilinear problems (5.32) and (5.48). Those estimates allow us to use the corresponding approximations for applied problems that model physical (or biological) processes in thick multilevel junctions of the type 3:2:2, especially at a neighborhood of the joint zone Ω'.

The obtained results are consistent with the results of Chaps. 2 and 4, namely, the main terms of the asymptotic approximations are solutions to the similar homogenized problems.

Asymptotic estimates (5.35) and (5.49) show us the influence of different components of the initial problems on the order of residuals from the asymptotic approximations. Summands $\|f_\varepsilon - f_0\|$ and $\delta_{\beta,1}\|g_\varepsilon - g_0\|$ correspond to discrepancies of the right-hand sides of the initial problem and homogenized one. In many textbooks and monographs on the asymptotic analysis, such kind of asymptotic estimates is typical (see e.g., [124, Chap. 2]). To formally improve estimates, in some articles there are special assumptions, namely, either the right-hand side is independent of ε, i.e., $f_\varepsilon \equiv f_0$, or $f_\varepsilon(x) = f_0(x) + \mathcal{O}(\varepsilon)$ as $\varepsilon \to 0$.

The summand $C_1\varepsilon^{1-\mu}$ bounds residuals left by the inner expansion. They are caused by the existence of non-energetic solutions with polynomial growth at infinity of junction-layer problems (5.18) and (5.19). As follows from Sect. 5.1.2, the behavior of such solutions is determined by the type $m : k : d$ of a thick junction. For instance, junction-layer non-energetic solutions have logarithmic growth at infinity for boundary-value problems in thick junctions of the type 3:1:1 (see [88, 109]).

The rest of the summands in (5.35) and (5.49) indicate the impact of the geometric structure of the thick junction and inhomogeneous perturbed Robin boundary conditions on the surfaces of the thin discs.

Boundary conditions on the bases of the cylinder Ω_0 can be replaced by boundary conditions of another type. But in the case of nonperiodic boundary conditions on $\partial \Omega_0 \cap \{r < r_0\}$ additional symmetry condition is needed, namely, the domain Π described in Sect. 5.1.2 has to be invariant with respect to the substitution $\xi_2 \mapsto 1 - \xi_2$.

Also in this chapter, we made special assumptions for the functions h_1 and h_2; they should be locally constant at a small enough neighborhood of the joint zone. This is a technical condition that allows to avoid additional bulky calculations.

References

1. Aiyappan, S., Nandakumaran, A.K.: Optimal control problem in a domain with branched structure and homogenization. Math. Methods Appl. Sci. **40**(8), 3173–3189 (2017)
2. Aiyappan, S., Nandakumaran, A.K., Prakash, R.: Generalization of unfolding operator for highly oscillating smooth boundary domains and homogenization. Calc. Var. Partial Differential Equations **57**(3), 86 (2018)
3. Aiyappan, S., Nandakumaran, A.K., Sufian, A.: Asymptotic analysis of a boundary optimal control problem on a general branched structure. Math. Methods Appl. Sci.
4. Amirat, Y., Bodart, O., De Maio, U., Gaudiello, A.: Asymptotic approximation of the solution of the Laplace equation in a domain with highly oscillating boundary. SIAM J. Math. Anal. **35**(6), 1598–1616 (2004)
5. Amirat, Y., Bodart, O., De Maio, U., Gaudiello, A.: Asymptotic approximation of the solution of Stokes equations in a domain with highly oscillating boundary. Ann. Univ. Ferrara Sez. VII Sci. Mat. **53**(2), 135–148 (2007)
6. Amirat, Y., Bodart, O., De Maio, U., Gaudiello, A.: Effective boundary condition for Stokes flow over a very rough surface. J. Differential Equations **254**(8), 3395–3430 (2013)
7. Babuška, I., Výborný, R.: Continuous dependence of eigenvalues on the domain. Czechoslovak Math. J. **15 (90)**, 169–178 (1965)
8. Benkaddour, A., Sanchez-Hubert, J.: Spectral study of a coupled compact-noncompact problem. RAIRO Modél. Math. Anal. Numér. **26**, 659–672 (1992)
9. Blanchard, D., Carbone, L., Gaudiello, A.: Homogenization of a monotone problem in a domain with oscillating boundary. M2AN Math. Model. Numer. Anal. **33**(5), 1057–1070 (1999)
10. Blanchard, D., Gaudiello, A.: Homogenization of highly oscillating boundaries and reduction of dimension for a monotone problem. ESAIM Control Optim. Calc. Var. **9**, 449–460 (2003)
11. Blanchard, D., Gaudiello, A., Griso, G.: Junction of a periodic family of elastic rods with a 3d plate. I. J. Math. Pures Appl. (9) **88**(1), 1–33 (2007)
12. Blanchard, D., Gaudiello, A., Griso, G.: Junction of a periodic family of elastic rods with a thin plate. II. J. Math. Pures Appl. (9) **88**(2), 149–190 (2007)
13. Blanchard, D., Gaudiello, A., Mel'nyk, T.A.: Boundary homogenization and reduction of dimension in a Kirchhoff-Love plate. SIAM J. Math. Anal. **39**(6), 1764–1787 (2008)
14. Blanchard, D., Gaudiello, A., Mossino, J.: Highly oscillating boundaries and reduction of dimension: the critical case. Anal. Appl. (Singap.) **5**(2), 137–163 (2007)
15. Brizzi, R., Chalot, J.P.: Boundary homogenization and Neumann boundary value problem. Ricerche Mat. **46**(2), 341–387 (1997)

16. Chechkin, G.A., Chechkina, T.P., D'Apice, C., De Maio, U., Mel'nyk, T.A.: Asymptotic analysis of a boundary-value problem in a cascade thick junction with a random transmission zone. Applicable Analysis **88**(10-11), 1543–1562 (2009)
17. Chechkin, G.A., Chechkina, T.P., D'Apice, C., De Maio, U., Mel'nyk, T.A.: Homogenization of 3d thick cascade junction with a random transmission zone periodic in one direction. Russian Journal of Mathematical Physics **17**(1), 35–55 (2010)
18. Chechkin, G.A., Mel'nyk, T.A.: Asymptotics of eigenelements to spectral problem in thick cascade junction with concentrated masses. Appl. Anal. **91**(6), 1055–1095 (2012)
19. Chechkin, G.A., Mel'nyk, T.A.: High-frequency cell vibrations and spatial skin effect in thick cascade junction with heavy concentrated masses. Comptes Rendus Mecanique **342**, 221–228 (2014)
20. Chechkin, G.A., Mel'nyk, T.A.: Spatial-skin effect for eigenvibrations of a thick cascade junction with 'heavy' concentrated masses. Math. Methods Appl. Sci. **37**(1), 56–74 (2014)
21. Cioranescu, D., Damlamian, A., Griso, G.: Periodic unfolding and homogenization. C. R. Math. Acad. Sci. Paris **335**(1), 99–104 (2002)
22. Cioranescu, D., Paulin, J.S.J.: Homogenization in open sets with holes. J. Math. Anal. Appl. **71**, 590–607 (1979)
23. Conca, C., Díaz, J.I., Liñán, A., Timofte, C.: Homogenization in chemical reactive flows. Electron. J. Differential Equations pp. No. 40, 22 pp. (electronic) (2004)
24. Corbo Esposito, A., Donato, P., Gaudiello, A., Picard, C.: Homogenization of the p-Laplacian in a domain with oscillating boundary. Comm. Appl. Nonlinear Anal. **4**(4), 1–23 (1997)
25. Craighead, H.G.: Nanoelectromechanical systems. Science **290**, 1532–1535 (2000)
26. D'Apice, C., De Maio, U., Mel'nyk, T.A.: Asymptotic analysis of a perturbed parabolic problem in a thick junction of type 3:2:2. Netw. Heterog. Media **2**(2), 255–277 (2007)
27. De Maio, U., Durante, T., Mel'nyk, T.A.: Asymptotic approximation for the solution to the Robin problem in a thick multi-level junction. Math. Models Methods Appl. Sci. **15**(12), 1897–1921 (2005)
28. De Maio, U., Gaudiello, A., Lefter, C.: Optimal control for a parabolic problem in a domain with highly oscillating boundary. Appl. Anal. **83**(12), 1245–1264 (2004)
29. De Maio, U., Mel'nyk, T.A.: Asymptotic analysis of the Neumann problem for the Ukawa equation in a thick multi-structure of type 3:2:2. In: Elliptic and parabolic problems, *Progr. Nonlinear Differential Equations Appl.*, vol. 63, pp. 207–215. Birkhäuser, Basel (2005)
30. De Maio, U., Mel'nyk, T.A.: Asymptotic solution to a mixed boundary-value problem in a thick multi-structure of type 3:2:2. Ukr. Math. Bull. **2**(4), 467–485 (2005)
31. De Maio, U., Mel'nyk, T.A.: Homogenization of the Neumann problem in thick multi-structures of type 3:2:2. Math. Methods Appl. Sci. **28**(7), 865–879 (2005)
32. De Maio, U., Mel'nyk, T.A.: Homogenization of the Robin problem for the Poisson equation in a thick multi-structure of type 3:2:2. Asympt. Anal. **41**(2), 161–177 (2005)
33. De Maio, U., Mel'nyk, T.A., Perugia, C.: Homogenization of the robin problem in a thick multilevel junction. Nonlinear Oscillations **7**(3), 326–345 (2004)
34. De Maio, U., Nandakumaran, A.K., Perugia, C.: Exact internal controllability for the wave equation in a domain with oscillating boundary with Neumann boundary condition. Evol. Equ. Control Theory **4**(3), 325–346 (2015)
35. Durante, T., Faella, L., Perugia, C.: Homogenization and behaviour of optimal controls for the wave equation in domains with oscillating boundary. NoDEA Nonlinear Differential Equations Appl. **14**(5-6), 455–489 (2007)
36. Durante, T., Mel'nyk, T.A.: Asymptotic analysis of a parabolic problem in a thick two-level junction. Zh. Mat. Fiz. Anal. Geom. **3**(3), 313–341 (2007)
37. Durante, T., Mel'nyk, T.A.: Asymptotic analysis of an optimal control problem involving a thick two-level junction with alternate type of controls. J. Optim. Theory Appl. **144**(2), 205–225 (2010)
38. Durante, T., Mel'nyk, T.A.: Homogenization of quasilinear optimal control problems involving a thick multilevel junction of type 3:2:1. ESAIM Control Optim. Calc. Var. **18**(2), 583–610 (2012)

39. Durante, T., Mel'nyk, T.A., Vashchuk, P.S.: Asymptotic approximation for the solution to a boundary-value problem with varying type of boundary conditions in a thick two-level junction. Nonlinear oscillations **9**(3), 326–345 (2006)
40. Feng, Z.C.: Handbook of zinc oxide and related materials: volume two, devices and nano-engineering, vol. 2. CRC press (2012)
41. Fleury, F., Sánchez-Palencia, E.: Asymptotics and spectral properties of the acoustic vibrations of a body perforated by narrow channels. Bull. Sci. Math. (2) **110**, 149–176 (1986)
42. Gajewski, H., Gröger, K., Zacharias, K.: Nonlinear operator equations and operator differential equations. Mir, Moscow (1978)
43. Gaudiello, A.: Asymptotic behaviour of non-homogeneous Neumann problems in domains with oscillating boundary. Ricerche Mat. **43**(2), 239–292 (1994)
44. Gaudiello, A., Guibé, O.: Homogenization of an elliptic second-order problem with $L \log L$ data in a domain with oscillating boundary. Commun. Contemp. Math. **15**(6), 1350008, 13 (2013)
45. Gaudiello, A., Guibé, O.: Homogenization of an evolution problem with $L \log L$ data in a domain with oscillating boundary. Ann. Mat. Pura Appl. (4) **197**(1), 153–169 (2018)
46. Gaudiello, A., Guibé, O., Murat, F.: Homogenization of the brush problem with a source term in L^1. Arch. Ration. Mech. Anal. **225**(1), 1–64 (2017)
47. Gaudiello, A., Hadiji, R., Picard, C.: Homogenization of the Ginzburg-Landau equation in a domain with oscillating boundary. Commun. Appl. Anal. **7**(2–3), 209–223 (2003)
48. Gaudiello, A., Lenczner, M.: A two-dimensional electrostatic model of interdigitated comb drive in longitudinal mode. Preprint arXiv:1906.01872v1 [math.AP] 5 Jun 2019
49. Gaudiello, A., Mel'nyk, T.A.: Homogenization of a nonlinear monotone problem with non-linear signorini boundary conditions in a domain with highly rough boundary. Journal of Differential Equations **265**, 5419–5454 (2018)
50. Gaudiello, A., Mel'nyk, T.A.: Homogenization of a nonlinear monotone problem with a big nonlinear Signorini boundary interaction in a domain with highly rough boundary. Nonlinearity **32**(12), 5150–5169 (2019). https://doi.org/10.1088/1361-6544/ab46e9
51. Griniv, R., Mel'nyk, T.: On the singular rayleigh functional. Mathematical Notes **60**(1), 97–100 (1996)
52. Grisvard, P.: Elliptic problems in nonsmooth domains. Boston, MA (1985)
53. Hempel, R., Post, O.: Spectral gaps for periodic elliptic operators with high contrast: an overview. In: Progress in analysis. Vol. I, II. Proceedings of the 3rd international congress of the International Society for Analysis, its Applications and Computation (ISAAC), Berlin, Germany, August 20–25, 2001, pp. 577–587. River Edge, NJ: World Scientific (2003)
54. Il'in, A.M.: A boundary value problem for an elliptic equation of second order in a domain with a narrow slit. I. The two-dimensional case. Mat. Sb. (N.S.) **99(141)**(4), 514–537 (1976)
55. Il'in, A.M.: Matching of asymptotic expansions of solutions of boundary value problems. Providence, RI: American Mathematical Society (1992)
56. Il'inskii, A.S., Slepyan, G.Y.: Oscillations and waves in electrodynamic systems with losses. Moscow State Univ, Moscow (1983)
57. Jimbo, S.: The singularly perturbed domain and the characterization for the eigenfunctions with Neumann boundary condition. J. Differential Equations **77**(2), 322–350 (1989)
58. Kazmerchuk, Y.A., Mel'nyk, T.A.: Homogenization of the Signorini boundary-value problem in a thick plane junction. Nelīnīĭnī Koliv. **12**(1), 44–58 (2009)
59. Kesavan, S., Saint Jean Paulin, J.: Optimal control on perforated domains. J. Math. Anal. Appl. **229**, 563–586 (1999)
60. Khruslov, E.Y.: On resonance phenomena in a diffraction problem (in Russian). Teor. Funkts. Funkts. Anal. Prilozhen. **10**, 113–120 (1968)
61. Kogut, P., Mel'nyk, T.A.: Asymptotic analysis of optimal control problems in thick multi-structures. IFAC Proceedings Volumes **37**(17), 265–275 (2004)
62. Kogut, P.I., Mel'nyk, T.A.: Limit analysis of a class of optimal control problems in thick singular junctions. J. of Automat. and Infor. Sci. **37**(1), 8–23 (2005)

63. Kondrat'ev, V.A., Oleinik, O.A.: Boundary-value problems for partial differential equations in non-smooth domains. Russian Mathematical Surveys **38**(2), 1–86 (1983)
64. Kotlyarov, V.P., Khruslov, E.Y.: On a limit boundary condition of some Neumann problem (in Russian). Teor. Funkts. Funkts. Anal. Prilozhen. **10**, 83–96 (1970)
65. Kovalevskii, A.: On the γ-convergence of integral functionals defined on sobolev weakly connected spaces. Ukrainian Mathematical Journal **48**, 683–698 (1996)
66. Kozlov, V.A., Maz'ya, V.G., Rossmann, J.: Elliptic boundary value problems in domains with point singularities, *Mathematical Surveys and Monographs*, vol. 52. Providence, RI: American Mathematical Society (AMS) (1997)
67. Ladyzhenskaya, O.A.: The boundary value problems of mathematical physics, *Applied Mathematical Sciences*, vol. 49. Springer-Verlag, New York (1985)
68. Ladyzhenskaya, O.A., Solonnikov, V., Uraltseva, N.: Linear and Quasi-linear Equations of Parabolic Type. American Mathematical Society, Providence, RI (1968)
69. Lenczner, M.: Multiscale model for atomic force microscope array mechanical behavior. Appl. Phys. Lett. **90**, 901–908 (2007)
70. Lions, J.L.: Some Methods of Solving Nonlinear Boundary Value Problems. Dunod-Gauthier-Villars, Paris (1969)
71. Lyshevski, S.E.: MEMS and NEMS: systems, devices, and structures. CRC press (2002)
72. Mahadevan, R., Nandakumaran, A.K., Prakash, R.: Homogenization of an elliptic equation in a domain with oscillating boundary with non-homogeneous non-linear boundary conditions. Applied Mathematics & Optimization (2018)
73. Marchenko, V.A., Khruslov, E.Y.: Boundary-value problems in domains with fine-grained boundary (in Russian). Naukova Dumka, Kiev (1974)
74. Marchenko, V.A., Khruslov, E.Y.: Homogenization of partial differential equations, *Progress in Mathematical Physics*, vol. 46. Birkhäuser Boston, Inc., Boston, MA (2006)
75. Mel'nyk, T.A.: Spectral properties of the discontinuous self-adjoint operator-functions. Dokl. Nats. Akad. Nauk of Ukraine **12**, 33–36 (1994)
76. Mel'nyk, T.A.: Asymptotic expansions of eigenvalues and eigenfunctions for elliptic boundary-value problems with rapidly oscillating coefficients in a perforated cube. J Math Sci **75**(3), 1646–1671 (1995)
77. Mel'nyk, T.A.: The asymptotics of the spectrum of the Dirichlet problem in a domain of fine tooth comb type. Russian Mathematical Surveys **51**(5), 940–941 (1996)
78. Mel'nyk, T.A.: Asymptotic analysis of the spectral boundary-value problems in thick singularly degenerate junctions of the different types. In: Multiple Scale Analysis and Coupled Physical Systems, Proceedings of Saint-Venant Symposium, Paris, August 28-29, 1997, pp. 453–459. The Presses des Ponts et Chaussees (1997)
79. Mel'nyk, T.A.: Homogenization of the Poisson equation in a thick periodic junction. Z. Anal. Anwendungen **18**(4), 953–975 (1999)
80. Mel'nyk, T.A.: On free vibrations of a thick periodic junction with concentrated masses on the fine rods. Nonlinear Oscillations **2**(4), 511–523 (1999)
81. Mel'nyk, T.A.: Asymptotic analysis of a spectral problem in a periodic thick junction of type 3:2:1. Math. Methods Appl. Sci. **23**(4), 321–346 (2000)
82. Mel'nyk, T.A.: Asymptotics of the eigenvalues and eigenfunctions of a boundary value problem in a thick periodic junction of type 3: 2: 2. Bulletin of the University of Lviv, Series: Mathematics **58**, 153–160 (2000)
83. Mel'nyk, T.A.: Homogenization of a singularly perturbed parabolic problem in a thick periodic junction of type 3:2:1. Ukraïn. Mat. Zh. **52**(11), 1524–1533 (2000)
84. Mel'nyk, T.A.: Asymptotic behavior of eigenvalues and eigenfunctions of the Steklov problem in a thick periodic junction. Nonlinear Oscil. **4**(1), 91–105 (2001)
85. Mel'nyk, T.A.: Asymptotic behaviour of eigenvalues and eigenfunctions for the Fourier problem in a thick junction of the type 3:2:1. In: Grupovi ta analitychni metody v matematichnij fizytsi, pp. 187–196. Kyïv: Instytut Matematyky NAN Ukraïny (2001)
86. Mel'nyk, T.A.: Vibrations of a thick periodic junction with concentrated masses. Math. Models Methods Appl. Sci. **11**(6), 1001–1027 (2001)

87. Mel'nyk, T.A.: Eigenmodes and pseudo-eigenmodes of thick multi-level junctions. In: Proceedings of the International Conference Days on Diffraction-2004 (St. Petersburg, June 29–July 2, 2004), pp. 51–52 (2004)
88. Mel'nyk, T.A.: Asymptotic analysis of the spectral Neumann problem in thick multi-structure of type 3:1:1. Nelineĭn. Granichnye Zadachi **15**, 85–98 (2005)
89. Mel'nyk, T.A.: Asymptotic behavior of eigenvalues and eigenfunctions of the fourier problem in a thick multilevel junction. Ukrainian Mathematical Journal **58**(2), 220–243 (2006)
90. Mel'nyk, T.A.: Homogenization of a boundary-value problem with a nonlinear boundary condition in a thick junction of type 3:2:1. Math. Methods Appl. Sci. **31**(9), 1005–1027 (2008)
91. Mel'nyk, T.A.: Asymptotic analysis of spectral problems in thick multi-level junctions. In: Integral methods in science and engineering. Vol. 1, pp. 205–215. Birkhäuser Boston, Inc., Boston, MA (2010)
92. Mel'nyk, T.A.: Asymptotic approximation for the solution to a semi-linear parabolic problem in a thick junction with the branched structure. J. Math. Anal. Appl. **424**(2), 1237–1260 (2015)
93. Mel'nyk, T.A.: Asymptotic approximations for chemical reactive flows in thick fractal junctions. In: Integral methods in science and engineering, pp. 387–399. Birkhäuser/Springer, Cham (2015)
94. Mel'nyk, T.A.: Asymptotic analysis of a mathematical model of the atherosclerosis development. International Journal of Biomathematics **12**(2, 1950014), 26 (2019)
95. Mel'nyk, T.A., Chechkin, G.A.: Asymptotic analysis of boundary value problems in thick cascade junctions. Report of the National Academy of Sciences of Ukraine, (9), 16–22 (2008)
96. Mel'nyk, T.A., Chechkin, G.A.: Homogenization of a boundary value problem in a thick cascade junction. J. Math. Sci. (N.Y.) **154**(1), 50–77 (2008)
97. Mel'nyk, T.A., Chechkin, G.A.: Asymptotic analysis of boundary value problems in thick three-dimensional multilevel junctions. Sb. Mat. **200**(3), 357–383 (2009)
98. Mel'nyk, T.A., Chechkin, G.A.: On new types of vibrations of thick cascade junctions with concentrated masses. Doklady Mathematics **87**(1), 102–106 (2013)
99. Mel'nyk, T.A., Chechkin, G.A.: Eigenvibrations of thick cascade junctions with 'very heavy' concentrated masses. Izvestiya: Mathematics **79**(3), 467–516 (2015)
100. Mel'nyk, T.A., Chechkin, G.A., Chechkina, T.P.: Convergence theorems for solutions and energy functionals of boundary value problems in thick multilevel junctions of a new type with perturbed Neumann conditions on the boundary of thin rectangles. J. Math. Sci. (N.Y.) **159**(1), 113–132 (2009)
101. Mel'nyk, T.A., Nakvasiuk, I.A.: Homogenization of a parabolic Signorini boundary value problem in a thick plane junction. J. Math. Sci. (N.Y.) **181**(5), 613–631 (2012). Problems in mathematical analysis. No. 62
102. Mel'nyk, T.A., Nakvasiuk, I.A.: Homogenization of a semilinear variational inequality in a thick multi-level junction. J. Inequal. Appl. **104**, 22 (2016)
103. Mel'nyk, T.A., Nakvasiuk, I.A., Wendland, W.L.: Homogenization of the Signorini boundary-value problem in a thick junction and boundary integral equations for the homogenized problem. Math. Methods Appl. Sci. **34**(7), 758–775 (2011)
104. Mel'nyk, T.A., Nazarov, S.A.: Asymptotic structure of spectrum of the Neumann problem in domains of dense-comb type. Russian Mathematical Surveys **48**(4), 228–229 (1993)
105. Mel'nyk, T.A., Nazarov, S.A.: The asymptotic structure of the spectrum in the problem of harmonic oscillations of a hub with heavy spokes. Russian Acad. Sci. Dokl. Math. **48**(3), 428–432 (1994)
106. Mel'nyk, T.A., Nazarov, S.A.: Asymptotic structure of the spectrum of the Neumann problem in a thin comb-like domain. C. R. Acad. Sci. Paris Sér. I Math. **319**(12), 1343–1348 (1994)
107. Mel'nyk, T.A., Nazarov, S.A.: Asymptotics of Neumann problem eigenvalues in a dense comb type domain. Russian Acad. Sci. Dokl. Math. **51**(3), 326–328 (1995)
108. Mel'nyk, T.A., Nazarov, S.A.: Asymptotics of the neumann spectral problem solution in a domain of thick comb type. J. Math. Sci. **85**(6), 2326–2346 (1997)

109. Mel'nyk, T.A., Nazarov, S.A.: Asymptotic analysis of the Neumann problem on the junction of a body and thin heavy rods. St. Petersbg. Math. J. **12**(2), 317–351 (2001)
110. Mel'nyk, T.A., Sadovyi, D.Yu.: Homogenization of elliptic problems with alternating boundary conditions in a thick two-level junction of type 3:2:2. J. Math. Sci. (N.Y.) **165**(1), 81–104 (2010)
111. Mel'nyk, T.A., Sadovyi, D.Yu.: Homogenization of a quasilinear parabolic problem with different alternating nonlinear Fourier boundary conditions in a two-level thick junction of the type 3:2:2. Ukrainian Math. J. **63**(12), 1855–1882 (2012)
112. Mel'nyk, T.A., Sadovyi, D.Yu.: Homogenization of boundary value problems in two-level thick junctions consisting of thin disks with rounded or sharp edges. J. Math. Sci. (N.Y.) **191**(2), 254–279 (2013)
113. Mel'nyk, T.A., Sadovyj, D.Yu.: Homogenization of quasilinear parabolic problems with alternating nonlinear Fourier and uniform Dirichlet boundary conditions in a thick two-level junction of type 3:2:2. Mat. visnyk NTSh **7**, 115–136 (2011)
114. Mel'nyk, T.A., Vashchuk, P.S.: Homogenization of a boundary-value problem with varying type of boundary conditions in a thick two-level junction. Nonlinear oscillations **8**(2), 240–255 (2005)
115. Mel'nyk, T.A., Vashchuk, P.S.: Homogenization of a boundary value problem with boundary conditions of mixed type in a thick junction. Diff. eq. **43**(5), 696–703 (2007)
116. Mikhailov, V.P.: Partial Differential Equations. Mir, Moscow (1983)
117. Mossino, J., Sili, A.: Limit behavior of thin heterogeneous domain with rapidly oscillating boundary. Ric. Mat. **56**(1), 119–148 (2007)
118. Murat, F., Tartar, L.: H-convergence. In: A. Cherkaev, R. Kohn (eds.) Topics in the mathematical modelling of composite materials, *Progress in Nonlinear Differential Equations and their Applications*, vol. 31, pp. 21–44. Birkhäuser Boston, Inc., Boston, MA (1997)
119. Nandakumaran, A.K., Prakash, R., Sardar, B.C.: Periodic controls in an oscillating domain: controls via unfolding and homogenization. SIAM J. Control Optim. **53**(5), 3245–3269 (2015)
120. Nandakumaran, A.K., Prakash, R., Sardar, B.C.: Asymptotic analysis of Neumann periodic optimal boundary control problem. Math. Methods Appl. Sci. **39**(15), 4354–4374 (2016)
121. Nazarov, S., Taskinen, J.: Asymptotics of the solution to the Neumann problem in a thin domain with sharp edge. Journal of Mathematical Sciences **142**(6), 2630–2644 (2007)
122. Nazarov, S.A., Plamenevsky, B.A.: Elliptic Problems in Domains with Piecewise Smooth Boundaries. De Gruyter (2011)
123. Nazarov, S.A., Taskinen, Y.: Asymptotic behavior of the solution of the Neumann problem in a thin domain with a sharp edge. Zap. Nauchn. Sem. S.-Peterburg. Otdel. Mat. Inst. Steklov. (POMI) **332**, 193–219, 317–318 (2006)
124. Oleinik, O.A., Shamaev, A.S., Yosifian, G.A.: Mathematical problems in elasticity and homogenization, vol. 26. Elsevier (1992)
125. Pao, C.V.: Nonlinear Parabolic and Elliptic Equations. Plenum Press: New York (1992)
126. Prakash, R., Sardar, B.C.: Homogenization of boundary optimal control problem in a domain with highly oscillating boundary via periodic unfolding method. Nonlinear Stud. **22**(2), 213–240 (2015)
127. Sadovyi, D.Yu.: Asymptotic approximation of solution to quasilinear elliptic boundary-value problem in a two-level thick junction of type 3:2:2. Carpathian Math. Publ. **4**(2), 297–315 (2012)
128. Sadovyj, D.Yu.: Asymptotic approximation of solution to quasilinear parabolic boundary-value problem in a two-level thick junction of type 3:2:2. Mat. Studii **38**(1), 51–65 (2012)
129. Sadovyj, D.Yu.: Asymptotic approximations of solutions to elliptic and parabolic boundary-value problems in a two-level thick junction of type 3:2:2. Adv. Appl. Math. Sci. **11**(8), 381–413 (2012)
130. Sanchez Hubert, J., Sánchez-Palencia, E.: Vibration and coupling of continuous systems. Springer-Verlag, Berlin (1989). Asymptotic methods
131. Showalter, R.E.: Monotone operators in Banach space and nonlinear partial differential equations. Providence, RI: American Mathematical Soiety (1997)

132. Taylor, M.E.: Partial Differential Equations III, *Appl. Math. Sci.*, vol. 117. Springer (1996)
133. Vishik, M.I.: Boundary problems for elliptic equations degenerating on the boundary of a region. Mat. Sb. N.S. **35(77)**, 513–568 (1954)
134. Zhikov, V., Kozlov, S., Olejnik, O.: Homogenization of differential operators and integral functionals. Transl. from the Russian by G. A. Yosifian. Berlin: Springer-Verlag (1994)

Index

© The Author(s), under exclusive license to Springer Nature Switzerland AG 2019
T. Mel'nyk and D. Sadovyi, *Multiple-Scale Analysis of Boundary-Value Problems in Thick Multi-Level Junctions of Type 3:2:2*, SpringerBriefs in Mathematics, https://doi.org/10.1007/978-3-030-35537-1

Printed in the United States
By Bookmasters